Springer-Verlag Berlin Heidelberg GmbH

Reviews of

142 Physiology Biochemistry and Pharmacology

Editors

M.P. Blaustein, Baltimore   R. Greger, Freiburg
H. Grunicke, Innsbruck   R. Jahn, Göttingen
W.J. Lederer, Baltimore   L.M. Mendell, Stony Brook
A.Miyajima, Tokyo   N. Pfanner, Freiburg
G. Schultz, Berlin   M. Schweiger, Berlin

With 11 Figures and 3 Tables

 Springer

ISSN 0303-4240
ISBN 978-3-662-31012-0   ISBN 978-3-540-44451-0 (eBook)
DOI 10.1007/978-3-540-44451-0
Library of Congress-Catalog-Card Number 74-3674

© Springer -Verlag Berlin Heidelberg 2001
Originally published by Springer-Verlag Berlin Heidelberg New York in 2001.
Softcover reprint of the hardcover 1st edition 2001

Production: PRO EDIT GmbH, 69126 Heidelberg, Germany
Printed on acid-free paper – SPIN: 10718053   27/3136wg-5 4 3 2 1 0

Contents

Indexed in Current Contents

Contents

Modulation of Protein Kinase C in Antitumor Treatment

J. Hofmann

Institute of Medical Chemistry and Biochemistry University of Innsbruck
A-6020 Innsbruck Austria

Contents

1 Introduction

Protein kinase C (PKC) is a family of serine/threonine specific protein kinases. The PKC isoenzymes can be classified into three groups: i) the conventional (cPKCs) α, βI, βII, and γ (require negatively charged phospholipids, diacylglycerol or phorbol ester, and calcium for optimal activation), ii) the novel (nPKCs) δ, ε, θ, η/L (mouse/human) and μ (require negatively charged phospholipids, diacylglycerol or phorbol ester, but no calcium), and iii) the atypical (aPKCs) λ/ι (mouse/human) and ζ (do not require calcium, diacylglycerol or phorbol ester, but only negatively charged phospholipids for optimal activity) (Nishizuka, 1995; Newton and Johnson, 1998). The PKC isoenzymes (Fig. 1) are characterized by four conserved (C1–C4) and five variable (V1–V5) domains (Stabel and Parker, 1991; Azzi et al., 1992; Hug and Sarre, 1993; Stabel, 1994). The regulatory domain consists of the C1 and the C2 region. C1 contains the pseudosubstrate region that can inhibit the enzyme by binding to the catalytic site (C4). In PKCμ, the pseudosubstrate domain is lacking. C1 also contains tandemly repeated cysteine-rich regions to which DAG (diacylglycerol), phorbol esters and bryostatins can bind. cPKCs and nPKCs contain two zinc fingers in the phorbol ester binding site, aPKCs are characterized by a single zinc finger. C2 contains the calcium binding region present only in cPKCs but not in nPKCs and aPKCs. Between the C2 and the C3 region the so called hinge region is situated which serves as cleavage site for calpain and trypsin during degradation. The C3 region is believed to be the ATP binding site and the C4 region the catalytic site.

PKC isoenzymes seem to play an important role in activation of signal transduction pathways leading to synaptic transmissions, the activation of ion fluxes, secretion, proliferation, cell cycle control, differentiation or tumorigenesis. PKC has become of major interest as target for therapeutic

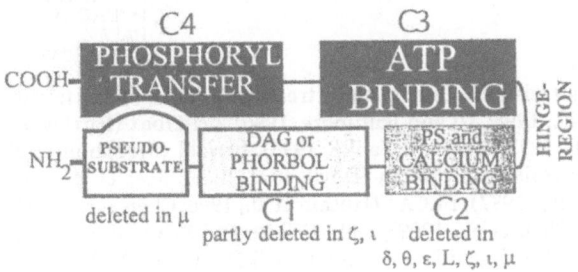

Fig. 1. Schematic representation of protein kinase C. The cartoon shows the different domains of PKC in the inactive form of the enzyme

J. Hofmann

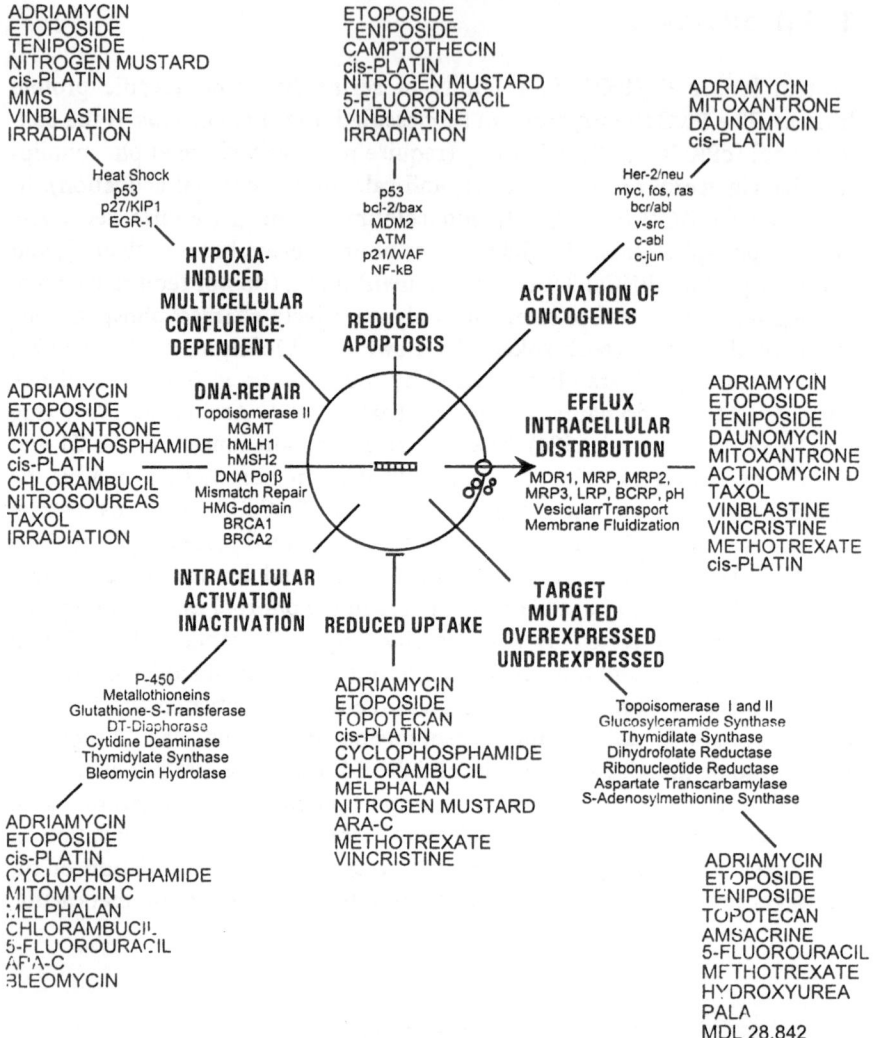

Fig. 2. Mechanisms causing resistance to antitumor treatment. ATM, ataxia telangiectasia gene, (Westphal et al., 1998; Xu and Baltimore, 1996), bcl-2/bax (Farrow and Brown, 1996, Zunino et al., 1997; Haq and Zanke, 1998), bcr/abl (McGahon et al., 1994), BCRP, breast cancer resistance protein (Doyle et al., 1998; Ross et al, 1999); bleomycin hydrolase (El-Deiry, 1997), BRCA1 (Husain et al., 1998; Chen et al., 1998), BRCA2 (Chen et al., 1998; Chen et al., 1999), c-abl (White and Prives, 1999), c-jun (Sanchez-Perez and Perona, 1999), cytidine deaminase (El-Deiry, 1997), DNA polβ, DNA polymerase β (Ochs et al., 1999), dihydrofolate reductase (Schimke, 1986), DT-diaphorase (Riley and Workman, 1992; Fitzsimmons et al., 1996; El-Deiry, 1997), EGR-1 (Ahmed et al., 1996), fos (Niimi et al., 1991), glucosylceramide synthase

intervention in a range of different diseases (Gescher et al., 1992; Bradshaw et al., 1993; Basu 1993; Deacon et al., 1997; Nixon, 1997; Goekjian and Jirousek, 1999). PKC may be involved in chronic granulomatous disease, allergy, asthma, rheumatoid arthritis (Westmacott et al, 1991), transplantation (Woodley et al., 1991), AIDS (Kinter et al., 1990; Accornero et al., 1998), Alzheimer's disease (Chauhan et al., 1991), multiple sclerosis (Defranco, 1991), hypertension (Ek et al., 1989), cardiac hypertrophy (Kwiatkowska-Patzer and Domanska-Janik, 1991), atherosclerosis (Kariya et al., 1987), diabetes (Inoguchi et al., 1992; Ishii et al., 1996) and cancer (Basu, 1993; Blobe et al., 1994, Gescher, 1998). PKC is the intracellular receptor for tumor promoting phorbol esters (Castagna et al., 1982; Niedel et al., 1983; Leach et al., 1983). Short term exposure of intact cells with phorbol esters such as 12-O-tetradecanoylphorbol-13-acetate (phorbol-12-myristate-13-acetate, TPA) activates PKC, long term exposure down-regulates PKC activity. Phorbol esters are able to promote tumor formation. Therefore, it was presumed that activation of PKC by TPA induces tumors and inhibition may reduce carcinogenesis or inhibit tumor growth. However, investigations revealed that the situation is more complicated. For example, bryostatin 1, another PKC

(Lavie et al., 1997; Liu et al., 1999), glutathione-S-transferase (Ozols et al., 1990, Tew,1994), heat shock (Ciocca et al., 1993); her-2/neu (Tsai et al., 1996), HMG-domains (Huang et al., 1994), hMLH1 (De las Alas et al. 1997), hMSH2 (Aebi et al., 1996), hypoxia (Bush et al., 1978; Vaupel et al., 1989; Sakata et al., 1991; Höckel et al., 1996), LRP, lung resistance-related protein (Scheffer et al., 1995), MDL 28,842, (Z)-5'-fluoro-4',5'-didehydro-5'-deoxyadenosine (Dwivedi et al., 1999), mdm2 (Kondo et al., 1995a), MDR1 (Juliano and Ling, 1976; Gottesman and Pastan, 1993), membrane fluidization (Regev et al., 1999), metallothioneins (Kelley et al., 1988; Kaina et al., 1990; Kondo et al., 1995b), MGMT, O^6-methylguanine-DNA methyltransferase (Erickson, 1991; Mattern et al., 1998), methyl methanesulfonate (MMS; Chen et al., 1998), mismatch repair (Moreland et al., 1999, White and Prives, 1999), MRP, multidrug resistance-associated protein (Cole et al., 1994), MRP2 (Cui et al., 1999; Hooijberg et al., 1999), MRP3 (Kool et al., 1999), multicellular resistance (Sutherland, 1988; Kobayashi et al., 1993; Pizao et al., 1993; Graham et al., 1994; StCroix et al., 1996), myc (Sklar and Prochownik, 1991), NF-κB (Wang et al., 1999b), P-450 (Doehmer et al., 1993), p53 (Lowe et al., 1993; Levine, 1997; Zunino et al., 1997; Piovesan et al., 1998); p27/KIP (StCroix et al., 1996); p21/WAF1 (Wang and Walsh, 1996; McDonald et al., 1996; Fan et al., 1997), pH (Martinez-Zaguilan et al, 1999; Williams et al., 1999), ras (Scanlon et al., 1991; Isonishi et al., 1991; El-Deiry, 1997), reduced uptake (Fry and Jackson, 1986; Perez et al., 1990; Slapak et al., 1990; Chu, 1994; Ma et al., 1998a; Moscow, 1998; Ma et al., 1998b), repair (Masumoto et al., 1999; Chen et al., 1998; Husain et al., 1998, Chen et al., 1999), ribonucleotide reductase (Ask et al, 1993; Yen et al., 1994), thymidilate synthase (Kinsella et al., 1997), topoisomerase I and II (Yarbro, 1992; Robert and Larsen, 1998), vesicular transport (Dietel et al., 1990), v-src (Masumoto et al., 1999)

modulator with properties similar to those of TPA (Blumberg, 1991; Kennedy et al., 1992; Szallasi et al., 1994) does not induce tumor formation. The compound exhibits potent antitumor activity and is currently undergoing phase I (Jayson et al., 1995, Grant et al., 1998) and phase II (Propper et al., 1998) clinical evaluation as an anticancer drug.

In addition to be a direct target of antitumor treatment, PKC has been shown to be involved in the resistance to antitumor treatment and in the modulation of apoptosis. Resistance to cancer chemotherapy is a major

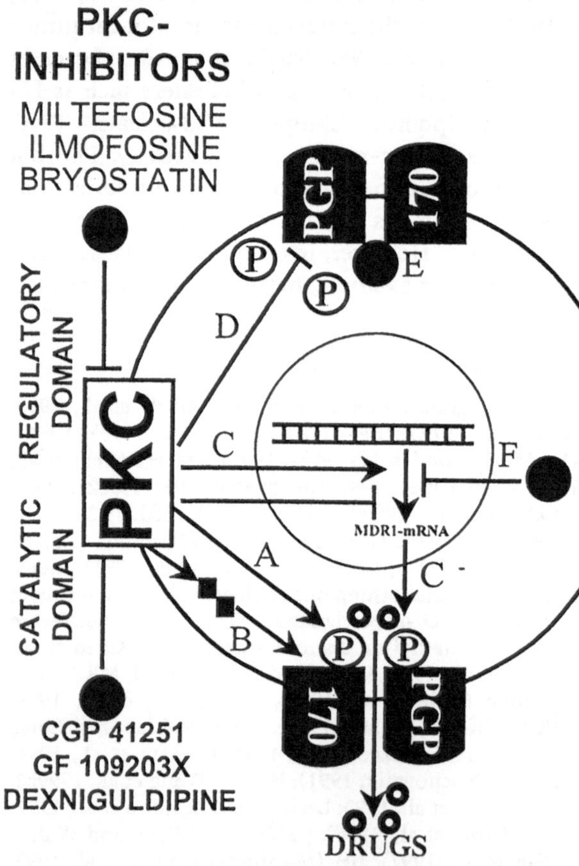

PKC-INHIBITORS

MILTEFOSINE
ILMOFOSINE
BRYOSTATIN

REGULATORY DOMAIN

CATALYTIC DOMAIN

PGP

170

P

P

E

D

C

F

MDR1-mRNA

A

C -

B

P P

170 PGP

CGP 41251
GF 109203X
DEXNIGULDIPINE

DRUGS

Fig. 3. Possible interactions between PKC and MDR1-mediated drug resistance. Activation of PKC might activate the drug efflux by phosphorylation of PGP (A), induce or activate proteins which modulate PGP (B, Castro et al., 1999), or induce the transcription and translation of MDR1-mRNA (C). Inhibitors of PKC might prevent phosphorylation of PGP leading to a decrease the drug efflux (D), inhibit the efflux of drugs by direct interaction with the drug binding site(s) or the ATP-binding sites of PGP (E), or prevent the expression of MDR1-mRNA (F)

problem in the treatment of cancer (Goldie and Coldman, 1984). A variety of mechanisms causing resistance have been observed (Fig. 2), among them: enhanced DNA-repair, reduced drug uptake, intracellular inactivation of drugs, reduced prodrug activation, mutated, overexpressed or not expressed targets, hypoxia, cell-cell interactions (multicellular resistance), increased drug efflux by overexpression of multidrug resistance gene 1 (MDR1), multidrug resistance-associated gene (MRP) or lung resistance protein (LRP), export of drugs by vesicles, intracellular compartimentalization of drugs, and altered expression or mutation of genes involved in apoptosis. Members of the PKC family have been reported be involved in multidrug resistance (MDR) and in apoptosis. Figure 3 shows an overview of the possible mechanisms by which PKC might interfere with MDR1-mediated drug resistance. However, the results of investigations concerning PKC and resistance are contradicting. This review summarizes data indicating a role of PKC in antitumor treatment and apoptosis, and on the other hand, data that argue against an involvement of PKC. The association of experimental results with pro or contra of a contribution of PKC is not always unambiguous.

2 PKC in Cell Proliferation and Tumor Growth

2.1 PKC Isoenzymes and Cell Proliferation

Investigations into the expression of distinct PKC isoenzymes in various tissues revealed a highly variable tissue distribution. PKCα and PKCζ are ubiquitously expressed. Brain contains all isoenzymes, whereas others such as skin and skeletal muscle contain only a few (Blobe et al., 1994). Such a different pattern of expression suggests that the PKC isoenzymes play different roles in the tissue of expression and do not suggest a general role of all isoenzymes in cell proliferation. In many publications an influence of PKC on cell proliferation has been reported.

Deletion or mutation of PKC1, the only member of the PKC family expressed in Saccharomyces cerevisiae (Mellor and Parker, 1998) led to osmotic instability, an arrest of protein synthesis and cell proliferation (Levin et al., 1990; Levin and Bartrett-Heubusch, 1992). NIH3T3 cells overexpressing PKCα exhibited an altered, "transformed-like" morphology, an increased growth rate, a higher saturation density and were able to grow in soft agar after treatment with TPA. These effects could be reversed by the unspecific PKC inhibitor staurosporine (Finkenzeller et al., 1992). With respect to untreated control liver, an activation and increased expression of PKCα was observed in diethylnitrosamine-induced liver tumors and lung metastases (La Porta et al., 1997). Early events in the transformation of keratinocytes have been found to be the mutation and activation of ras, activation of the epidermal growth factor receptor, upregulation of PKCα, inactivation of through tyrosine phosphorylation (Yuspa, 1998), and downregulation of the expression of PKCδ (Geiges et al., 1995). In cultured myoblasts PKCα was found to have an important role in maintaining proliferation (Capiati et al., 1999). PKCα has been shown to activate telomerase in human breast cancer cells which may represent an essential step in the maintenance of proliferation in human cancers (Li et al., 1998). On the other hand, increased expression of PKCα led to cessation of growth, induction of differentiation in B16 melanoma cells and to gene dose-dependent inhibition of proliferation in K562 cells (Gruber et al., 1992). Recombinant chimaeras with the regulatory domain of PKCα and catalytic domains of other PKC isoenzymes inhibited cell growth (Acs et al., 1997). PKCα overexpressing bovine aortic endothelial cells exhibited reduced proliferation and increased accumulation in the G2/M phase of the cell cycle (Rosales et al., 1998). Overexpression of PKCα in MCF10A cells suppressed proliferation endowing cells with properties consistent with a metastatic phenotype (Sun and Rotenberg, 1999).

PKCβ overexpressing cells were more susceptible to transformation with the H-ras oncogene (Hsiao et al., 1989). Overexpression of PKCβI in rat fibroblasts led to disordered growth (Housey et al., 1988). The human erythroleukemia K562 cell line expresses PKCα, βII and ζ. The cells undergo megakaryocytic differentiation and cessation of proliferation when treated with TPA. K562 cells overexpressing human PKCα grew more slowly and were more sensitive to the cytostatic effects of TPA than control cells, whereas cells overexpressing βII were less sensitive to TPA. Antisense experiments demonstrated that PKCβII is required for K562 cell proliferation, whereas PKCα is involved in megakaryocytic differentiation (Murray et al., 1993). In F9 embryonal carcinoma PKCα seems to play an active role in differentiation and PKCβ activity is incompatible with differentiation (Cho et al., 1998). PKCβI overexpressing bovine aortic endothelial cells promoted growth and shortened the doubling time, whereas PKCα exhibited reduced proliferation (Rosales et al., 1998). However, in human HL-60 promyelocytes activation of PKCβ was found to be necessary and sufficient for TPA-induced differentiation (MacFarlane and Manzel, 1994). In murine keratinocytes overexpression of βI led to growth inhibition and Ca^{2+}-induced differentiation (O'Driscoll et al., 1994). PKCβI expression was higher in a well-differentiated SKUT-1-B mixed mesodermal uterine cell line compared to the moderately differentiated endometrial HEC-1-B adenocarcinoma cell line (Bamberger et al., 1996). As shown in mice homozygous for a targeted disruption of the PKCβII gene, this isoenzyme seems to play an important role in B-cell activation (Leitges et al., 1996).

When PKCγ was overexpressed in NIH 3T3 cells, reduced growth factor requirements, growth to higher saturation density and formation of tumors in nude mice were observed (Persons et al., 1988). Human U251 MG glioma cells overexpressing PKCγ showed an increased rate of growth in monolayer culture, increased colony-forming efficiency on soft agar, and increased DNA synthesis in response to epidermal growth factor and basic fibroblast growth factor (Mishima et al., 1994).

PKCδ seems to be involved in growth inhibition, differentiation, apoptosis and tumor suppression (Gschwendt, 1999). TPA stimulation of PKCδ overexpressing CHO cells led to a cell division arrest (Watanabe et al., 1992). PKC δ overexpressing glioma cells showed a decreased rate of growth and decreased colony-forming efficiency (Mishima et al, 1994). However, overexpression of PKCδ in rat HT mammary adenocarcinoma cell lines significantly increased anchorage-independent growth, although it had no effect on growth of adherent cells. PKCδ seems to be involved in regulating attachment and anchorage-independence, which may be related to increased metastatic potential in this system (Kiley et al., 1999). Recombinant chimaeras with the regulatory domain of PKCδ and catalytic domains of other PKC isoenzymes inhibited cell growth (Acs et al., 1997). Proteolytic cleavage of PKCδ was found to activate this isoenzyme during apoptosis (Emoto et al., 1995; Ghayur et al., 1996). The antitumor agents taxol, vinblastine and vincristine specifically activated PKCδ (Das et al., 1998). Transgenic mice overexpressing PKCδ in the epidermis were found to be resistant to skin tumor promotion by TPA (Reddig et al., 1999). In NIH 3T3 cells which normally express only PKCα, overexpression of PKCδ by transfection induced significant changes in morphology and caused the cells to grow more slowly and to a decreased cell density in confluent cultures. These changes were accentuated by treatment with TPA. (Mischak et al., 1993). Overexpression of PKCε did not lead to morphological changes, but caused increased growth rates and higher cell densities in monolayers. None of the PKCδ overexpressers grew in soft agar with or without TPA, but all the cell lines that overexpressed PKCε grew in soft agar in the absence of TPA, but not in its presence. NIH 3T3 cells that overexpressed PKCε also formed tumors in nude mice with 100% incidence, indicating that high expression of PKCε contributes to neoplastic transformation (Mischak et al., 1993). Overexpression of PKCε in Rat 6 embryo fibroblasts led to a 7–13-fold increase in Ca^{2+}-independent PKC activity, to formation of dense foci in monolayer culture, decreased doubling time, increased saturation density, decreased serum requirement, growth in soft agar, and tumor formation in nude mice (Cacace et al., 1993). In nontumorigenic rat colonic epithelial cells overexpression of PKCε caused marked morphological changes in two transfected

clones, which were accompanied by increased saturation densities and anchorage-independent colony formation in semisolid agar. These growth
effects were attenuated or reversed by chronic incubation with TPA (Perletti
et al., 1996). A small cell lung cancer cell line that exhibited rapid growth
compared to other small cell lung cancer cell lines overexpressed a constitutively active catalytic fragment of PKCε (Baxter et al., 1992). Dominant
negative PKCε inhibited the proliferation of NIH3T3 cells. Constitutively
active PKCα or PKCε overcame this inhibitory effect (Cai et al., 1997). Recombinant chimaeras between the regulatory domains of PKCα, PKCδ,
PKCε and the catalytic domains of these isoenzymes were transfected into
NIH 3T3 cells. All chimaeras containing a regulatory or a catalytic domain of
PKCε exhibited growth-promoting activity (Acs et al., 1997). These data
indicate that PKCε seems to have oncogenic properties. However, it has also
been reported, that inhibition of cell proliferation by tamoxifen is associated
with activation of PKCε (Lavie et al., 1998). In neuronal cells PKCε seems to
be important for differentiation (Ohmichi et al., 1993; Hundle et al., 1995;
Fagerstrom et al., 1996).

Overexpression of PKCζ has been reported to be required for mitogenic
maturation of Xenopus oocytes and led to deregulation of growth control in
mouse fibroblasts (Berra et al., 1993). However, these effects of PKCζ in
Xenopus oocytes seem not to be clear (Carnero et al., 1995). In U937 monocytic leukemia cells PKCζ overexpression decreased proliferation rate and
saturation density, indicating the induction of differentiation (Ways et al.,
1994). In normal NIH3T3 fibroblasts (Montaner et al., 1995; Crespo et al.,
1995) and K562 cells (Murray et al., 1997) no effects of PKCζ overexpression
on cell proliferation or oncogenic transformation were observed. PKCζ
overexpression in v-raf-transformed NIH-3T3 cells drastically retarded
proliferation, abolished anchorage-independent growth, and reverted the
morphological transformation (Kieser et al., 1996). Activation of PKCζ by
ceramide in acute lymphoblastic leukemia MOLT-4 cells induced apoptosis.
However, it was also shown that ceramide treatment, in addition to activation of PKCζ, inactivated PKCα (Lee et al, 1996a). Exposure of cells to a
genotoxic stimulus that induced apoptosis, led to an inhibition of PKCζ
(Berra et al, 1997). The product of the par-4 gene interacted with PKCζ and
inhibited its enzymatic activity. The expression of par-4 correlated with
growth inhibition and apoptosis (Diaz-Meco et al, 1996). In Cos-7 cells conventional and novel PKCs activated the ERK/MAPK cascade via raf-1,
whereas PKCζ stimulated this pathway without raf-1 activation (Schönwasser et al., 1998). N-myc acts to increase the malignancy of neuroblastoma
cells. Overexpression of N-myc in these cells caused suppression of PKCδ
and induction of PKCζ (Bernards, 1991).

PKCμ seems to be involved in murine keratinocyte proliferation. A correlation between PKCμ expression and enhanced cell proliferation was also observed for NIH3T3 mouse fibroblasts overexpressing human PKCμ (Rennecke et al., 1999).

2.2 PKC Expression in Tumor Cells and Tumors

Overexpression of PKC seems to be involved in breast cancers. Elevated levels of PKC activity in breast tumors relative to normal breast tissue was found by O'Brian et al. (1989a). TPA inhibited the growth of mammary carcinoma MCF-7, BT-20, MDA-MB-231, ZR-75-1, and HBL-100, but not that of T-47-D cells. TPA-non-responsive T-47-D cells exhibited the lowest PKC activity. A rapid TPA-dependent translocation of cytosolic PKC to membranes was found in the five TPA-sensitive cell lines without affecting cell growth. However, TPA-treatment for more than 10 hours inhibited reversibly the growth of TPA-responsive cells. This effect coincided with the complete loss of cellular PKC activity due to the proteolysis of the translocated membrane-bound PKC. Resumption of cell growth after TPA-removal was closely related to the specific reappearance of the PKC activity in the TPA-responsive human mammary tumor cell lines suggesting an involvement of PKC in growth regulation (Fabbro et al., 1986). During a 4-day culture period, various phorbol ester derivatives inhibited the proliferation of MCF-7 breast carcinoma cells in a dose-dependent manner. A correlation between the relative potencies of the various phorbol ester derivatives for inhibiting both phorbol-12,13-dibutyrate (PdBu) binding and cell proliferation was found (Darbon et al., 1986). However, it was also reported that PKC activation by TPA and DAG inhibited MCF-7 cell proliferation (Issandou and Darbon, 1988; Issandou et al., 1988). PKC activity was described to be higher in estrogen receptor negative human mammary tumor cells compared to estrogen-receptor-containing counterparts (Borner et al., 1987; Ways et al., 1995; Morse-Gaudio et al., 1998). MCF-7 cells transfected with PKCα displayed an enhanced proliferative rate, anchorage-independent growth, dramatic morphologic alterations including loss of an epithelioid appearance, and increased tumorigenicity in nude mice. PKCα overexpressing MCF-7 cells exhibited a significant reduction in estrogen receptor expression and decreases in estrogen-dependent gene expression (Ways et al., 1995). Phorbol esters were found to down-regulate the expression of estrogen receptors in breast cancer cell lines (Hähnel and Gschwendt, 1995). PKCα was found to be activated in situ in a significant number of human breast tumors (Ng et al., 1999). MCF-7 breast cancer cells transfected with PKCα led to a more aggressive phenotype compared to untransfected cells (Ways et al., 1995).

Differential display between the non-metastatic and the PKCα overexpressing metastatic MCF-7 cells showed that a homologue to a putative glialblastoma cell differentiation-related protein is upregulated. Histone 3B, integrins 3α and 6α were downregulated (Carey and Noti, 1999). In contrast to the publication by Ways et al. (1995), another report showed that overexpression of PKCα in MCF-7 cells caused upregulation of PKCβ and this led to a less aggressive phenotype, which was characterized by reduced in vitro invasiveness and markedly diminished tumor formation and growth in nude mice. These findings were explained by the down-regulation of estrogen receptor levels observed in tumors derived from PKCα-infected MCF-7 cells (Manni et al., 1996). TPA and bryostatin 1 inhibited the growth of MCF-7 breast cancer cells. TPA induced rapid translocation of PKCα protein and PKC activity to the membrane fraction of MCF-7 cells. In contrast, bryostatin 1 treatment resulted in the loss of the PKCα activity from both cytosolic and membrane compartments within 10 minutes of treatment. These results suggested that PKCα may specifically play a role in inhibiting growth of human breast cancer cells by bryostatin (Kennedy et al., 1992).

PKCα expression in human astrocytomas was found to be highest in well-differentiated (grade 1) tumors, intermediate in anaplastic (grade 2) astrocytomas, and low or nondetectable in dedifferentiated glioblastomas (grade 3 astrocytomas) and normal controls (Benzil et al., 1992). It was also found that the levels of PKCα in eight glioblastoma cells lines were similar to those in normal glial cells (Misra-Press et al., 1992). PKC activity was significantly higher in glioma cell lines compared to bladder, colorectal, rhabdomyosarcoma-oligodendrocyte hybrid, melanoma, cervix, and fibroblast cells, even though 3 of 8 of the non-glioma lines had higher proliferation rates than A172 glioma cells. In non-glioma cell lines, no correlation was found between the PKC activity and proliferation rates (Baltuch et al., 1993). In a comparison between rat C6 glioma cells and non-malignant rat astrocytes, both C6 glioma cells and astrocytes were found to express PKCα, β, δ, ε and ζ, but not γ. Enzyme activity measurements revealed that the elevated PKC activity of glioma cells was due to overexpression of PKCα (Baltuch et al., 1995). In another investigation all human glioma lines examined, and the rat glioma C6, displayed high PKC activity relative to nonmalignant glial cells, which correlated with their proliferation rates over their respective growth phase. Frozen surgical human malignant glioma specimens also displayed high PKC activity (Couldwell et al., 1992). The administration of PdBu or TPA resulted in a dose-related inhibition of growth of human glioma cell lines in vitro. The synthetic nonphorbol PKC activator SC-9 produced an even more pronounced decrease in cell proliferation. Conversely, the administration of 4-αTPA, a phorbol ester that binds but does not activate

PKC, had no effect on the proliferation rate. In contrast to the response of glioma cells, nonmalignant human adult astrocytes treated with the PKC activators responded by increasing their proliferation rate. The opposed effects of PKC activators on nonmalignant astrocytes versus glioma growth may be due to a high intrinsic PKC activity in glioma cells, with resultant down-regulation of enzyme activity following the administration of the pharmacological activators (Couldwell et al., 1990). Cell lines derived from high-grade gliomas expressed higher levels of PKCα than did cell lines derived from low-grade gliomas. In glioblastoma-derived cell lines PKCα was mainly expressed in the cytosolic fraction, indicating an inactive state of the enzyme. Bryostatin 1 specifically down-regulated PKCα in glioblastoma-derived cell lines. However, this was not associated with significant growth inhibition illustrating that PKCα seems not to be essential for proliferation (Zellner et al., 1998). Compared to two normal glial cell lines, in eight glioblastoma cell lines PKCα and PKCγ were similar. PKCε was elevated three to thirty times in six of the eight tumors. PKCζ was elevated twofold in all of the tumors (Xiao et al., 1994).

Elevated levels of PKC were also found in thyroid cancers if compared to normal thyroid tissue (Hagiwara et al., 1990; Hatada et al., 1992). TPA enhanced, tamoxifen and staurosporine inhibited invasion and growth of estrogen receptor-negative follicular thyroid cancer cells (Hoelting et al., 1996). In human thyroid cancers (Prevostel et al., 1995; Prevostel et al., 1997) and in pituitary cancers (Alvaro et al., 1997; Alvaro et al., 1993) point mutations in PKCα were detected. However, in 11 human pituitary tumours cDNA was subcloned and up to ten individual clones were sequenced from each tumour, resulting in 85 clones analyzed in total. All of the pituitary adenomas showed a normal wild-type sequence of PKCα DNA. Even if the tumor was invasive (infiltration of the dura mater) no mutation was found. Moreover, Western blot analyses did not show any differences in PKCα protein expression in invasive as compared with noninvasive pituitary adenomas. These data argue against suggestions that mutated PKCα is a feature of invasive pituitary tumours (Schiemann et al., 1997). PKC activity and expression were higher in adenomatous pituitaries than in normal human or rat pituitaries. PKC expression in growth hormone-secreting and non-secreting tumors was significantly higher than that in prolactin-secreting tumors. PKC activity was significantly higher in invasive tumors than in non-invasive tumors. In 3 adenomas which were obtained from patients treated with bromocriptine or octreotide, particulate- and soluble-PKC activities were significantly lower than those measured in non-treated adenomas (Alvaro et al., 1992). In human pituitary tumors predominantly PKCα in all adenomas, and variable expression of PKCβ and PKCγ in some tumors was found. Normal and neoplastic pituitaries expressed abundant mRNA for PKCε, whereas some tumors and one normal pituitary had a few cells positive for PKCζ (Jin et al., 1993).

A murine UV-induced fibrosarcoma cell line had an unusual PKC subcellular distribution with 87% of the PKC activity associated with the membrane. Sequencing of PKCα DNA from ultraviolet-induced-fibrosarcoma cells showed four point mutations in the fibrosarcoma PKC, of which three are in the highly conserved regulatory domain and one is in the conserved region of the catalytic domain. Expression of this mutant PKCα gene in normal Balb/c 3T3 fibroblasts resulted in a fibrosarcoma-like PKC membrane localization and in cell transformation, as judged by their formation of dense foci, anchorage-independent growth and ability to induce solid tumours when inoculated into nude mice. By contrast, transfectants expressing the normal PKCα cDNA did not display a morphology typical of malignant transformed cells and failed to induce tumours in vivo. These findings seemed to demonstrate that point mutations in the primary structure of PKC modulate enzyme function and are responsible for inducing oncogenicity (Megidish and Mazurek, 1989). However, these results could not be reproduced (Borner et al., 1991). The nontumorigenic, immortal line of murine melanocytes, Mel-ab, required the continual presence of biologically active phorbol esters for growth. Comparable treatments of murine B16 melanoma cells resulted in partial inhibition of cell proliferation. Significant levels of PKC were present in quiescent Mel-ab cells, whereas no immunoreactive protein was detected in cell extracts from either proliferating Mel-ab or B16 cells. These data showed that PKC down-regulation, and not activation, correlates with the growth of melanocytes in culture (Wilson et al., 1989; Brooks et al., 1991). It has been shown that TPA stimulated the proliferation of normal human melanocytes, whereas it inhibited the growth of human melanoma cell lines. PKCδ, ε and ζ were detected in both normal melanocytes and in four melanoma cell lines. In contrast, both PKCα and β were expressed in normal melanocytes, whereas only either PKCα or β was detected in melanoma cells. TPA inhibited the growth of cells lacking PKCα more efficiently than the other melanoma cell lines which lacked PKCβ. It was further shown that PKCβ was not detected in freshly isolated human primary or metastatic melanoma tissues. (Oka et al., 1996). This may be a consequence of lack of induction of terminal differentiation by PKCβ. In human early prostatic adenocarcinomas an increase of PKCα, PKCε, PKCζ and a decrease of PKCβ was consistently observed during the genesis and progression of prostate cancer compared with nonneoplastic prostate tissues (Cornford et al., 1999). An increased nuclear PKCβ activity was observed in lung metastases compared to the parental liver tumor induced by diethylnitrosamine (La Porta et al., 1997). Spontaneously and chemically transformed mouse pulmonary epithelial cells exhibited reduced levels of PKC (Morris and Smith, 1992).

Underexpression of PKC seems to be involved in colon cancer. PKC has been shown to be important for growth arrest and differentiation of intestinal cells (Saxon et al., 1994, Assert et al., 1993). Human Vaco 10 MS colon cancer cells were growth-inhibited by activation of PKC (McBain et al., 1990). Significantly higher Ca^{2+}-dependent PKC activities were observed in both the cytosolic and particulate fractions of the normal mucosa relative to the corresponding values obtained with the human colon carcinoma fractions. The average specific activity ratios were 5.1 (normal cytosolic/carcinoma cytosolic) and 3.7 (normal particulate/carcinoma particulate) for PKC (Guillem et al., 1987). In patients with colonic adenomas and colonic carcinomas, total PKC activity was found to be significantly reduced as compared to adjacent mucosa (Kopp et al., 1991). In 15 of 15 primary human colon tumors there was a decrease of approximately 40% in the levels of diacylglycerol when compared to paired adjacent normal mucosa samples. Assays on the same samples indicated that this decrease was seen both in tumors that did and did not display mutations in codon 12 of c-K-ras. These results suggested that the PKC signal transduction pathway is suppressed in human colon cancer (Phan et al., 1991). In another investigation the mean value for cellular PKC enzyme activity in the colon tumors from 39 patients was approximately 60 percent of that found in the paired adjacent normal mucosa samples (Levy et al., 1993). In 18 human colonic adenomas and carcinomas a significant decrease in particulate PKC activity compared with the adjacent normal mucosa was observed. Decreased PKC activity correlated with increased adenoma size (Kusunoki et al., 1992). Five of six PKC isoenzymes present in normal mucosa showed reduced protein levels during tumor development in the human colon (Kahl-Rainer et al., 1994). Decrease of PKCα seems to be of major importance for development of colorectal cancers (Suga et al., 1998). DNAs for PKCα in sense or antisense orientations were transfected into human colonic adenocarcinoma CaCo-2 cells. Sense transfected clones exhibited 3-fold increases and antisense transfectants approximately 95% decreases in PKCα expression with no significant alterations in other PKC isoforms. Transfection of CaCo-2 cells with PKCα in the antisense orientation resulted in enhanced proliferation and decreased differentiation, as well as in a more aggressive transformed phenotype compared with empty vector-transfected control cells. In contrast, cells transfected with PKCα cDNA in the sense orientation demonstrated decreased proliferation, enhanced differentiation, and an attenuated tumor phenotype compared with these control cells. (Scaglione-Sewell et al., 1998). Human

HT-29M6 colon cancer cells transfected with an activated form of PKCα showed decreased proliferation and increased invasion due to alterations in cell adhesion. When these cells were xenografted into athymic mice, higher expression of activated PKCα led to a reduction in tumor size (Batlle et al., 1998). The abundance of PKCβ mRNAs was decreased in 30 of 39 colon tumors (Levy et al., 1993). HT-29 colon cancer cells transfected with rat PKCβI displayed increased doubling time, decreased saturation density, loss of anchorage-independent growth in soft agar, and decreased tumorigenicity in nude mice following exposure to TPA (Choi et al., 1990). Overexpression of PKCβI in SW480 colon cancer cells caused growth suppression (Goldstein et al., 1995). In small adenomas of 18 patients a significant increase of cytosolic PKCα and decrease of membrane PKCβII compared to normal neighboring mucosa was found. In 7 patients PKCδ was reduced (Assert et al., 1999). In human colon cancer tissue a significant decrease in PKCε was found (Pongracz et al., 1995a). Unsaturated free fatty acids and bile acids which are present in the colon are activators of PKC (Weinstein, 1991). It was reported that activation of PKC by unsaturated fatty acids was associated with proliferation of rat colonic epithelium (Craven and de Rubertis, 1988; Craven and de Rubertis, 1992).

2.3 Antiproliferative and Antitumor Effects of PKC-Modulation

2.3.1 Bryostatins

PKC is a phorbol ester receptor (Castagna et al., 1982, Niedel et al., 1983). Phorbol esters such as TPA or PdBu modulate PKC activity and, therefore, have been used for investigations into PKC effects on apoptosis and antitumor drug resistance. However, the use of TPA or PdBu can give apparently contradictory findings because they can activate several PKC isoenzymes. Unlike the physiological activator diacylglycerol, TPA is stable and induces a persistent activation leading to the downregulation of PKC (Blumberg, 1991). So short term exposure of intact cells to phorbol esters activates PKC, prolonged exposure depletes cells of the enzyme, most likely due to proteolytic degradation (Chida et al., 1986). In many cases it is difficult to distinguish whether the effects of phorbol esters are due to activation or inactivation of PKC. In addition to PKC, TPA also interacts with other proteins (Ahmed et al., 1993; Valverde et al., 1994; Caloca et al., 1997). Gnidimacrin is another PKC activator which leads to downmodulation of the enzyme following long term exposure. The compound exhibits antitumor activity in vivo (Yoshida et al., 1996).

Bryostatins are a group of macrocyclic lactones which are isolated from marine bryozoans (Pettit et al., 1970). These compounds exhibit a remarkable affinity for PKC where they compete with phorbol esters for the same binding site (Berkow and Kraft, 1989; Blumberg, 1991). The majority of investigations with bryostatins were carried out with bryostatin 1. In intact cells bryostatins, like phorbol esters, activate PKC and cause a translocation from the cytosol to the membrane (Hocevar and Fields, 1991). Following prolonged exposure, PKC is proteolytically degraded (Berkow and Kraft, 1989; Blumberg, 1991; Kennedy et al., 1992). Several differences between bryostatins and TPA have been observed. Like phorbol esters, bryostatins stimulate growth in some systems and are growth-inhibitory in others (Smith et al., 1985; Gschwendt et al., 1988; Kraft et al., 1988; Stone et al., 1988). However, in contrast to phorbol esters, bryostatins do not act as tumor promoters. In many systems bryostatin 1 induced only a subset of the responses to TPA and sometimes blocked those which it did not induce (Blumberg, 1991).

In NIH 3T3 fibroblasts bryostatin 1 showed similar potency to TPA for translocating PKCα to the cell membrane but was a much more potent downregulator of PKCα activity than TPA. It was also a much more potent translocator and downregulator of PKCδ and PKCε than TPA (Szallasi et al., 1994). The compound inhibited the proliferation of human A549 lung carcinoma, human MCF-7 breast cancer, murine renal adenocarcinoma, B16 melanoma, M5076 reticulum cell sarcoma, L10A B-cell lymphoma cells and exhibited antitumor activity in the murine P388 leukemia screening system (Dale and Gescher, 1989; Kennedy et al., 1992; Hornung et al., 1992). Another investigation showed that TPA and bryostatins 1 and 2 inhibited the growth of A549 cells. At high concentrations the bryostatins did not affect cell growth. Incubation with bryostatins 1 or 2 also led to PKC translocation, which was, however, much weaker than that observed with TPA. Exposure of cells to TPA or the bryostatins for longer than 30 minutes caused the gradual disappearance of total cellular PKC activity. PKC downregulation was concentration-dependent and complete after 24 hours. The bryostatins were potent inhibitors of the binding of [^{3}H]PdBu to its receptors in intact cells, and the inhibition was dependent on bryostatin concentration. (Dale et al., 1989). A P388 subline resistant to bryostatin was developed. The ability of the cytosol of these cells to phosphorylate PKC-specific substrate, phorbol ester binding and PKC isoenzyme expression was decreased. The resistant subline was also resistant to the activation by TPA (Prendiville et al., 1994). TPA inhibited growth of four out of six cell lines by up to 75% in 5-day cultures. Bryostatin 1 inhibited growth of MCF-7 cells only at a high dose. However, bryostatin 1 completely antagonized the growth inhibition and morphological changes induced by TPA in MCF-7 cells. The divergent effects of these two agents were associated with differing effects on PKC activity and isoform expression in MCF-7 cells. TPA induced rapid translocation of PKCα to the membrane fraction of MCF-7 cells. In contrast, bryostatin treatment resulted in the loss of PKCα from both cytosolic and membrane compartments within 10 minutes of treatment. In coincubation assays the bryostatin effect was dominant over that of TPA. Similar effects on PKCα were seen in the MDA-MB-468 cell line whose growth was inhibited by TPA but not by bryostatin 1. In contrast, in the T-47-D cell line, TPA was not growth inhibitory and failed to induce translocation of PKCα to the cell membrane. Bryostatin, however, still caused loss of PKCα and PKC activity from cytosolic and membrane fractions. Thus, differential actions of bryostatin 1 and TPA on PKC activity and the PKCα isoform level in the membrane-associated fraction of MCF-7 and MDA-MB-468 cells may account for the divergent effects of these two agents on cell growth and morphology (Kennedy et al., 1992). Bryostatin 1 produced faster than TPA, an inactive, dephosphorylated form of PKCα in the LLC-MK2 line of renal epithelial cells which predisposes the enzyme to proteolysis. This explains a more efficient downregulation of PKCα by bryostatin (Lee et al., 1996b). However, several experimental approaches suggested that the growth inhibition by bryostatin is independent of PKC. 26-epi-bryostatin 1 showed 60-fold reduced affinity for PKC and 30–60-fold reduced potency to translocate and downregulate PKC isozymes compared with bryostatin 1, but was similar potent in inhibiting the growth of B16/F10 mouse melanoma cells. 26-epi-bryostatin 1 was found to exhibit a 10-fold reduced toxicity in C57BL/6 mice compared with bryostatin 1. From these results it was concluded that the growth inhibition of the bryostatins, at least in this system, did not result from interaction with PKC (Szallasi et al., 1996).

In clinical phase I trials with bryostatin 1, myalgia appeared as the dose-limiting side effect. Conventional toxicities included fever, anemia, fatigue, phlebitis, headache and thrombopenia (Rea et al., 1992; Prendiville et al., 1993; Jayson et al., 1995; Grant et al., 1998). From 29 patients, 11 with relapsed non-Hodgkin's lymphoma and chronic lymphocytic leukemia achieved stable disease for 2 to 19 months. An in vitro assay for total PKC evaluation in the patient's peripheral-blood samples demonstrated activation within the first 2 hours with subsequent downregulation by 24 hours, which was maintained throughout the duration of the 72-hour infusion. (Varterasian et al., 1998). Of 15 patients with malignant melanoma pretreated with chemotherapy only one developed stable disease for 9 month, 14 developed progressive disease (Propper et al., 1998).

2.3.2 Phospholipid Analogues

Phospholipid analogues are a new class of drugs which exhibit broad antineoplastic activity (Berdel, 1991; Hilgard et al., 1993; Brachwitz and Vollgraf, 1995). Typical representatives from this group are 1-*O*-octadecyl-2-*O*-methyl-rac-glycero-3-phosphocholine (edelfosine, ET-18-OCH$_3$), ilmofosine (BM 41440) and hexadecylphosphocholine (miltefosine). These compounds exhibited remarkable antitumor activity against various experimental tumors in vitro as well as in vivo (Berdel, 1991; Workman, 1991). Miltefosine is the first of these compounds used in the clinic. In a phase II trial 8 (35%) partial remissions out of a total of 23 patients with metastatic breast cancer were observed (Clavel et al., 1992). The drug is approved in several countries for the topical treatment of skin metastases resulting from breast cancers (Hilgard et al., 1993). In the clinic an overall complete and partial response rate of 33%, and if clinically relevant minor responses were added, a total response rate of 53% could be observed (Klenner et al., 1998). The molecular mechanism responsible for the cytotoxic effect of these compounds is not quite clear. Interference with mitogenic signalling in the cell membrane seems to be the target of these compounds (Workman, 1991; Brachwitz and Vollgraf, 1995). The compounds inhibit PKC in cell free extracts by competing with phosphatidylserine (Kiss et al., 1987, Shoji et al., 1988, Hofmann et al., 1989, Shoji et al., 1991, Überall et al., 1991). They are equally as potent against PKC in intact cells as in cell free extracts (Shoji et al., 1988; Hofmann et al., 1989; Überall et al., 1991). The assumption that inhibition of PKC is responsible for the inhibition of cell growth is further supported by the observation that the antiproliferative effect of miltefosine can be antagonized by TPA (Geilen et al., 1991). Quaternary ammonium analogues of edelfosine, which inhibited PKC activity in vitro, also inhibited cell proliferation. A quaternary ammonium analogue which did not inhibit PKC did not inhibit cell proliferation (Civoli and Daniel, 1998). However, the conclusion that PKC is the target of phospholipid analogues has been questioned recently. Treatment of HL-60 and K562 cell lines with miltefosine and edelfosine for 2 hours did not lead to PKC translocation. In the same experimental setting, dioctanoylglycerol-stimulated PKC translocation was not affected by miltefosine or edelfosine. These findings indicated that miltefosine and edelfosine do not interfere with PKC translocation but rather mediate a general decrease of the enzyme activity in the membrane and cytosol of the cells. Since the extent of PKC inhibition was similar in the sensitive HL-60 and in the more resistant K562 cell line, it was concluded that inhibition of PKC may not be a prerequisite for the antiproliferative action of miltefosine and edelfosine (Berkovic et al., 1994). Cells depleted of PKC showed similar sensitivity to edelfosine on PKC activity in different cell types as cells expressing PKC. These results suggested that a role of PKC in the cytotoxic action of edelfosine is very unlikely (Heesbeen et al., 1994). These data may be explained by the fact that phospholipid analogues interfere with other signal transduction pathways as inositol phosphate formation (Überall et al., 1991; Powis et al., 1992) or phosphatidylcholine synthesis (Vogler et al., 1985).

Inhibition of cell proliferation of Chinese hamster ovary cells treated with the PKC inhibitor sphinganine correlated with PKC inhibition. Sphinganine blocked changes in protein phosphorylation patterns that occurred in response to TPA (and vice versa). Mutant cells that exhibited increased resistance to sphinganine lacked phorbol ester and sphinganine-induced phosphorylation changes and differed somewhat in the behavior of PKC in vitro. From these results it was concluded that the cytotoxicity of sphinganine may be a consequence of PKC inhibition (Stevens et al., 1990). N,N-dimethyl-sphingosine and N,N,N-trimethyl-D-erythro-sphinganine inhibited the proliferation of several human tumor cell lines in vitro and inhibited the growth of human gastric MKN74 carcinoma xenografts in athymic mice. Both compounds showed similar growth inhibitory effects, despite the fact that N,N,N-trimethyl-D-erythro-sphinganine showed a much stronger inhibitory effect than N,N-dimethylsphingosine on PKC activity (Endo et al., 1991).

2.3.3 Staurosporine-Derivatives

Staurosporine, isolated from Streptomyces (Omura et al., 1977) is a very potent inhibitor of PKC (Tamaoki et al., 1986). The compound is an unspecific kinase inhibitor and inhibits protein tyrosine kinases, cAMP-dependent kinases and myosin light-chain kinases with only moderately higher IC_{50} values than those for PKC (Meyer et al., 1989). Recently staurosporine-derivatives with higher specificity for PKC such as benzoyl-staurosporine (CGP 41251, Meyer et al., 1989; Caravatti et al., 1994; Marte et al., 1994), RO 31-8220 and RO 31-7549 (Dieter and Fitzke, 1991), Gö 6976 and Gö 6850 (Martiny-Baron et al., 1993; Hu, 1996), GF 109203X (Toullec et al., 1991), 7-hydroxstaurosporine (UCN-01, Seynaeve et al., 1994), CGP 53506 and CGP 54345 (Zimmermann et al., 1996), rottlerin (Gschwendt et al., 1994) and LY333531 (Ishii et al., 1996) have been synthesized. With the exception of rottlerin all of them exhibit the lowest IC_{50} values for cPKCs (Mizuno et al., 1995; Hofmann, 1997). Bryostatins and phospholipid analogues modulate PKC activity by interaction with the regulatory domain. Staurosporine and its derivatives are believed to inhibit PKC by competition with ATP.

CGP 41251 inhibits preferentially cPKCs (IC_{50} = 0.022-0.031 µM; Marte et al., 1994; Meggio et al., 1995; Fabbro et al., 1999). It also inhibits vascular endothelial growth factor(VEGF)-inducible KDR-receptor tyrosine kinase (IC_{50} = 0.086 µM). PDGF receptor tyrosine kinase is inhibited with an IC_{50} of 0.08 µM (Fabbro et al., 1999). CGP 41251 inhibited the proliferation of human bladder carcinoma, leukemia, small cell lung cancer, non-small cell lung cancer, glioblastoma, gliosarcoma and breast carcinoma cell lines (Meyer et al., 1989; Courage et al., 1995; Ikegami et al., 1995; Ikegami et al., 1996a; Begemann et al., 1996; Fabbro et al., 1999). The compound exhibited antitumor activity in human tumor xenografts derived from bladder, melanoma, gastric, colorectal, breast or lung cancers in nude mice (Meyer et al., 1989; Ikegami et al., 1995). Consistent with the finding that CGP 41251 blocks the VEGF-dependent phosphorylation of KDR (one of the two major receptors for VEGF) in vitro was the result obtained in mice that received a subcutaneous VEGF-impregnated implant. The treatment with CGP 41251 completely inhibited the angiogenic response to VEGF, but not to bFGF. Thus, CGP 41251 may suppress tumor growth by inhibiting tumor angiogenesis in addition to directly inhibiting tumor cell proliferation via its effects on PKCs. This anti-angiogenic action may contribute to the broad antitumor activity displayed by the compound (Fabbro et al., 1999). In clinical phase I (Czendlik and Graf, 1996; McDonald et al., 1997) and phase II studies (Mehta et al., 1997) CGP 41251 has been well tolerated.

UCN-01 has been shown to inhibit the proliferation of MCF-7, MDA-MB453, SK-BR-3 (Seynaeve et al., 1993), A549 and MCF-7 (Courage et al., 1995) and glioma cell lines (Pollack et al., 1996) in vitro. The compound inhibited the growth of myeloid leukemia, fibrosarcoma and epidermoid carcinoma xenografts in mice. UCN-01 inhibited the down-modulation of epidermal growth factor receptor caused by TPA in A431 cells at a near 50% inhibitory concentration for cell growth. (Akinaga et al., 1991). Clinical trials using the compound as an antitumor agent are ongoing (Lush et al., 1997). A major problem in the treatment of patients with both CGP 41251 and UCN-01, may be the binding to plasma proteins such as α-acidic protein (Lush et al., 1997; Sausville et al., 1998; Utz et al., 1998).

Several data indicate that PKC is the major target of these compounds, others do not support this notion. Inhibition of cell proliferation by CGP 41251 correlated with the inhibition of PKC. CGP 42700, a derivative of CGP 41251 did not inhibit PKC and did not inhibit cell proliferation (Meyer et al., 1989; Utz et al., 1998). Human A549 lung carcinoma cells were exposed to CGP 41251, UCN-01 or Ro 31-8220 at gradually increasing concentrations. Cells acquired, 4.3-fold resistance against CGP 41251, 4.0-fold against UCN-01 and 14-fold against Ro 31-8220. Cells were neither collaterally cross-resistant towards the PKC inhibitors nor resistant against the growth-inhibitory properties of TPA. PKC activity in these cells was decreased by between 57% and 96% compared to wild-type A549 cells. The levels of PKCα and PKCθ in all 3 resistant cell types and of PKCε in UCN-01-resistant cells were concomitantly reduced (Courage et al., 1997). Compounds with varying potencies for PKC inhibition were investigated for their antiproliferative activity in A549 and MCF-7 carcinoma cells. When the IC_{50} values for cell proliferation were plotted against the IC_{50} values for inhibition of cytosolic PKC activity, two groups of compounds could be distinguished. The group which comprised the more potent inhibitors of enzyme activity (calphostin C, staurosporine and its analogues UCN-01, RO 31-8220, CGP 41251) were the stronger growth inhibitors, whereas the weaker enzyme inhibitors (trimethylsphingosine, miltefosine, NPC-15437, H-7, H-7I) affected prolif-eration less potently. GF 109203X was exceptional in that it exhibited high potency for PKC inhibition but was only weakly cytostatic. The PKCα selec-tive inhibitor CGP 53506 has been shown to inhibit growth of T24 bladder carcinoma cells (Zimmermann et al, 1996). Novel diaminoanthraquinones also have therapeutical potentials against human tumors (Jiang et al., 1992).

In cells depleted of PKC by incubation with bryostatin, growth arrest in-duced by staurosporine, RO 31-8220, UCN-01 or H-7 was slightly, but not significantly lower than that observed in control cells not treated with bry-ostatin 1. These results suggested that PKC is unlikely to play a direct role in the arrest of the growth of A549 and MCF-7 cells mediated by these agents (Courage et al., 1995). The antiproliferative activity of these compounds may be caused at least in part by inhibition of kinases involved in cell cycle regulation as CDKs or retinoblastoma protein. It has been reported that CGP 41251 and UCN-01 interact with cell cycle progression (Begemann et al., 1996; Ikegami et al., 1996a; Ikegami et al., 1996b; Courage et al., 1996; Kaka-wami et al., 1996; Wang et al., 1996; Shimizu et al., 1996; Akiyama et al., 1997; Yu et al., 1998). CGP 41251 and UCN-01 are also involved in the induction of apoptosis (Wang et al., 1995; Han et al., 1996; Hunakova et al., 1996; Shao et al., 1997) which may be PKC-dependent or PKC-independent.

2.3.4 Antisense Oligonucleotides

Down-regulation of PKCα by antisense phosphorothioate oligonucleotides significantly reduced the rate of proliferation of three human glioma cell lines. This reduction in growth rate was attributed to apoptosis (Dooley et al., 1998). Expression of antisense oligonucleotides directed towards PKCα in U-87 glioblastoma cells resulted in no detectable PKCα content, a 95% reduction in total PKC activity, an increase in doubling time in vitro, less serum-dependent growth and reduced sensitivity to the selective PKC inhibitor Ro 31-8220 (Ahmad et al., 1994). Treatment of mice bearing U-87 xenografts with oligonucleotides against PKCα resulted in suppression of tumor growth (Yazaki et al., 1996). In T-24 human bladder carcinoma cells, expression of PKCα antisense led to a reduction of PKCα and was without effect on the expression of other PKC isoenzymes. The antisense oligonucleotide inhibited the growth of T-24 bladder carcinoma, A549 lung carcinoma and Colo 205 colon carcinoma cells in a dose-dependent manner in nude mice (Dean et al., 1996). The selective depletion of PKCα in smooth muscle cells by antisense towards PKCα inhibited proliferation, but did not induce apoptosis (Leszczynski et al., 1996). A 20-mer phosphorothioate oligodeoxynucleotide directed against human PKCα alone or in combination with established antitumor drugs were studied in nude mice that had been transplanted s.c. with a variety of human tumors (breast, prostate, large cell lung and small cell lung carcinomas, and melanomas). Additive antitumor effects with the antisense oligonucleotide and the cytotoxins were found for half of the combinations studied. The combination of the antisense oligonucleotides with vinblastine or cis-platin showed superadditive antitumor activities against MCF-7 human breast carcinomas and PC3 prostate carcinomas with complete responses, that with adriamycin resulted in superadditive antitumor effects against BT20 human breast carcinomas with complete tumor responses, and that with mitomycin C showed superadditive antitumor effects with cures observed against NCI-H460 human large cell carcinomas. The antisense oligonucleotide was completely inactive as single agent against A549 and NCI-H69 human lung carcinomas (Geiger et al., 1998). C8161 human melanoma cells were treated in vitro with a phosphorothioate antisense oligodeoxynucleotide that specifically inhibited PKCα. Northern blots demonstrated 70% inhibition of PKCα mRNA in treated cells compared to controls. Metastasis was suppressed by 75% when oligonucleotide-treated cells were injected intravenously into athymic mice (Dennis et al., 1998). The expression of antisense PKCα markedly inhibited the cell proliferation rate, colony forming efficiency in soft agar, and tumorigenicity of human LTEPa-2 lung carcinoma cells in nude mice (Wang et al., 1999a). Animals implanted with the pancreatic cancer cells overexpressing PKCα had a mortality rate almost twice that of those implanted with the parental non overexpressing cell line. Animals treated with antisense oligonucleotides directed against PKCα after orthotopic implantation of pancreatic cancer cells survived statistically longer than those treated with vehicle alone. Treatment with a scrambled oligonucleotide also conferred a survival benefit compared with vehicle alone (Denham et al., 1998). In a clinical phase I study with an antisense phosphorothioated oligonucleotide to PKCα, dose limiting was complement activation. From 36 patients with advanced cancer, two patients with non-Hodgkins's lympoma achieved complete remissions (Nemunaitis et al., 1999).

Antisense oligonucleotides directed against PKCβII led to loss of proliferative capacity in K562 cells (Murray et al., 1993). Downmodulation of PKCζ by antisense had no effect on the proliferation and saturation density in K562 cells (Murray et al., 1997). However, Spitaler et al. (1999) found that down-modulation of PKCζ with antisense oligonucleotides in HeLa cells led to the induction of apoptosis.

2.3.5 Remarks and Conclusions

PKC isoenzymes seem to play important roles in cell proliferation and tumor growth. There are several explanations for the contradicting results. Some experiments show that a certain PKC isoenzyme seems to be involved in proliferation, but other results indicate that this isoenzyme may be involved in differentiation and inhibition of proliferation. These highly variable effects may be due to the different cells used in these experiments. Many conclusions were made by overexpression of a certain PKC isoenzyme. In many experiments overexpression of PKC isozymes was induced by DNA-transfection. This may lead to overexpression at the "wrong time" or in the "wrong cell line". The localization of the PKC isoenzymes by receptors for active C-kinase (RACKs) or receptors for inactive C-kinase (RICKs) is important for the PKC activity and the phosphorylation of specific targets (Mochley-Rosen and Gordon, 1998). The levels of these proteins may depend on the tissue. The lack of such proteins may lead to inappropriate localization of PKC. It has been shown that mislocation of PKC isoenzymes led to altered behaviour of the cells to TPA (Whelan et al., 1999). Overexpression of a PKC isoenzyme does not necessarily lead to elevated PKC activity because the overexpressed protein may be inactive in the cytosol (Mischak et al., 1993). Following activation and downmodulation in physiological conditions, the catalytic domains of PKC isoenzymes may have some functions in the nuclear envelope, nuclear matrix, nucleolus or chromatin (Buchner, 1995; Martelli et al., 1999). In tumor cells increased expression of PKC isoenzymes may be important for tumor growth. However, the expression of PKC may also be accidental because of the unstability of the genome in tumor cells. It is not known exactly whether tumor promotion is the result of short term activation, prolonged activation or down-modulation. PKC modulators such as bryostatins, phospholipid analogues, UCN-01, CGP41251 or PKCα antisense show antitumor activity. However, it is not clear whether PKC is the only or the major target or whether other kinases are the main targets of these compounds. It was indeed presumed that growth inhibition by bryostatin does not result from interaction with PKC and that the interaction of bryostatin with PKC is responsible for the toxicity of the compound (Szallasi et al, 1996). Further investigations into the exact role of the different PKC isoenzymes in proliferation, differentiation and tumor growth are essential.

3 PKC in Apoptosis

3.1 Introduction

The control of cell numbers is regulated by cell proliferation, differentiation and apoptosis. Increased proliferation and/or decreased apoptosis result in neoplasia. In addition to inhibition of proliferation or induction of differentiation, the modulation of apoptosis can be employed for treatment for cancer. Several anticancer agents in use are potent inducers of apoptosis (Dive and Hickman, 1991; Fisher, 1994). Tumor promotion may result in decreased apoptosis. Because PKC activation by TPA induces carcinogenesis, it seems that PKC may be involved in apoptosis. There are many reports on the effects of PKC on apoptosis. However, the results are very controversial. Here an overview of these data is presented.

3.2 PKC Activation Promotes Apoptosis

Fragmentation of DNA induced by tyrosine-kinase inhibitors was enhanced by activation of PKC with TPA in mouse thymocytes (Azuma et al., 1993). Exposure of human myeloid leukemia HL-60 and KG-1 cells to mitoxantrone induced programmed cell death. Pretreatment with TPA enhanced mitoxantrone-induced apoptosis, whereas staurosporine and H-7 had no effect (Bhalla et al., 1993). Exposing HL-60 cells to TPA for 48 hours induced morphological changes characteristic of apoptosis. In contrast, TPA for five days did not induce apoptosis in a HL-60 mutant defective in its response to TPA. It was concluded that the promyelocytes have the capacity to undergo apoptosis in response to agents which activate PKC (MacFarlane and O'Donnell, 1993). In mouse but not in rat thymocytes, apoptosis was potentiated by TPA and prevented by H-7 (Shaposhnikova et al., 1994). In murine B lymphoma WEHI-231 cells signaling through anti-receptor antibodies led to growth arrest and apoptosis. Direct activation of PKC with phorbol esters also could mediate this response (Haggerty and Monroe, 1994). The deoxycytidine analogues ara-C and gemcitabine induced apoptosis in human ovarian BG-1 cancer cells and at the same time activated PKC. Short-term exposure to TPA and gemcitabine did not alter the response to gemcitabine. However, a 24-hour exposure to TPA followed by gemcitabine resulted in synergistic cytotoxicity. Coincubation of TPA with a PKC inhibitor abrogated the synergistic response (Cartee and Kucera, 1998, Cartee et al., 1998).

Overexpression of PKCζ in U937 histiocytic lymphoma cells increased expression of PKCα and β isoforms. In response to TPA, parental U937 cells displayed growth arrest and differentiated into a monocyte/macrophage-like cell line, while PKCζ overexpressing cells underwent death. The ability of GF109203X to inhibit TPA-induced cell death suggested that activation of a conventional isoform was necessary to induce apoptosis (de Vente et al., 1995). Activation of PKC enhanced and down-modulation of PKC activity reduced apoptosis of neuronal cells (Mailhos et al., 1994). Ceramide, an activator of PKCζ is an inducer of apoptosis (Muller et al, 1995; Hannun, 1998, Obeid et al., 1993).

3.3 PKC Activation Prevents Apoptosis

Activation of PKC by TPA suppressed IL-6-starvation-induced apoptosis in T1165 and T1198 plasmacytoma cells (Romanova et al., 1996). Stimulation of PKC by TPA blocked DNA fragmentation and cell death induced by the Calcium-ionophore A23187 or glucocorticoids in thymocytes (McConkey et al., 1989). A23187 was found to induce apoptosis in immature mouse thymocytes. The addition of TPA at low concentration inhibited the DNA fragmentation induced by A23187 and was accompanied by an increase in DNA synthesis. The result suggested that PKC activation switched a suicide process induced by A23187 to an opposite process (Kizaki et al., 1989). A combination of A23187 and TPA inhibited corticosterone-induced apoptosis in lymphocytes (Iseki et al., 1991). Usually chronic lymphocytic leukemia cells undergo apoptosis during culture for 1 to 2 days. DNA fragmentation was greatly enhanced when cells were cultured in the presence of colchicine, etoposide, or methylprednisolone. Phorbol esters inhibited cell death. Phorbol ester action was prevented by H-7 (Forbes et al., 1992). Activation of PKC by short term treatment of HL-60 cells with TPA inhibited apoptosis triggered by topoisomerase I and II inhibitors (Solary et al., 1993) and by singlet oxygen (Zhuang et al., 1998). Induction of apoptosis by C_2-ceramide or TNFα was prevented by TPA activation of PKC in U937 cells (Obeid et al., 1993). Activation of PKC by TPA also inhibited DNA damage-induced apoptosis in U937 cells (Kaneko et al., 1999). The activation of PKC promoted cell survival of mature lymphocytes prone to apoptosis and protected them from radiation-induced apoptosis (Lucas et al., 1994). Apoptosis occurring at a high rate among B cells in germinal centers can be arrested by

TPA (Knox et al., 1993). Activation of PKC by TPA prevented Fas-induced apoptosis in human leukemic T cell lines (Ruiz-Ruiz et al., 1997). Apoptosis of B cells occurring in the ileal Peyer's patch follicles in sheep could be abrogated during the first 12 hours of culture by the addition of TPA or PdBu (Motyka et al., 1993). Translocation of PKC from the cytosol rescued center B cells in germinal centers from apoptosis (Knox and Gordon, 1994).

Serum deprivation of C3H 10T½ fibroblasts resulted in DNA fragmentation which was prevented by TPA. Palmityl carnitine, an inhibitor of PKC, reversed the effects of TPA (Kanter et al., 1984). bFGF and phorbol esters protected endothelial cells against radiation-induced apoptosis. bFGF mediated the translocation of PKCα form the cytosol into the membrane (Haimovitz-Friedman, 1994a). PKC activation blocked radiation-induced apoptosis, and apoptosis was restored by ceramide analogues added exogenously (Haimovitz-Friedman et al., 1994b). Treatment of U937 and HL-60 cells with 0.5–1 μM daunorubicin induced a greater than 30% activation of neutral sphingomyelinase activity within 4–10 min with concomitant sphingomyelin hydrolysis and ceramide generation. Activation of PKC by TPA and phosphatidylserine inhibited daunorubicin-induced neutral sphingomyelinase activation, sphingomyelin hydrolysis, ceramide generation, and apoptosis. The apoptotic response could be restored by the addition of 25 μM C_6-ceramide. Therefore, it was concluded that PKC activity negatively regulates the anthracycline-activated sphingomyelin-ceramide apoptotic pathway (Mansat et al., 1997).

3.4 PKC Inhibition Promotes Apoptosis

Granulocyte macrophage colony-stimulating factor or interleukin-3 suppressed apoptosis in hemopoietic cells. H-7, staurosporine, and sphingosine reverted this suppression of apoptosis. Conversely, TPA allowed a bypass of receptor activation in suppression of apoptosis (Rajotte et al., 1992). Induction of apoptosis by freezing and rewarming of confluent human synovial McCoy's cells was inhibited by activation of PKC and promoted by H-7 or sphingosine (Perotti et al., 1990). H-7 enhanced dexamethasone induced apoptosis in mouse natural killer cells and cytotoxic T lymphocytes (Migliorati et al., 1994). Expression of v-abl prevented apoptosis in a haemopoietic cell line. Calphostin C restored apoptosis. Chronic exposure to TPA did not alter survival of the cells (Evans et al., 1995). Addition of chelerythrine or calphostin C to murine B lymphoma WEHI-231 cells triggered apoptosis (Chmura et al., 1996a). GF109203X reverted the suppression of apoptosis mediated by IL-2 or IL-2 plus dexamethasone in murine T-cells. The use of TPA allowed a bypass of the IL-2/IL-2R interaction in the suppression of apoptosis mediated by dexamethasone or IL-2 withdrawal in TS1 cells with medium but not in TS1 cells with high affinity IL-2 receptors. (Gomez et al., 1994). Bisindolylmaleimide VIII potentiated Fas-mediated apoptosis in human astrocytoma 1321N1 and in Molt-4 T cells, both of which were devoid of apoptosis induced by anti-Fas antibody in the absence of the PKC inhibitor (Zhou et al., 1999). Calphostin C and TNF-α were found to induce apoptosis in U937 histiocytic lymphoma cells. Treatment with TPA prevented apoptosis induced by these compounds. A peptide derived from the V1 region of PKCε specifically blocked translocation of PKCε and blocked also the inhibitory effect on apoptosis by TPA (Mayne and Murray, 1998). Staurosporine inhibition of PKC activity in lymphocytes correlated to some extent with the inhibition of [^{3}H]thymidine incorporation and the breakdown of DNA into oligonucleosome-sized fragments (Lucas et al., 1994). Calphostin C and chelerythrine induced apoptosis in HL-60 and U937 cells (Freemerman et al., 1996). In L1210 murine leukemia cell lines sensitive and resistant to apoptosis induced by cis-platin or 5-azacytidine, staurosporine induced apoptosis in both cell lines (Segal-Bendirdjian and Jacquemin-Sablon, 1996). Staurosporine and H-7 potentiated apoptosis triggered by singlet oxygen in HL-60 cells (Zhuang et al., 1998). Incubation of HL-60 cells with H-7, chelerythrine or calphostin C produced concentration-dependent increases in DNA fragmentation (Jarvis et al., 1994a). HL-60 cell transfected with bcl-2 were significantly less susceptible to apoptosis to ara-C-induced apoptosis than untransfected control cells. When bcl-2 over-expressing HL-60 cells were incubated with bryostatin, staurosporine or UCN-01, ara-C induced apoptosis was restored to levels greater than those observed in empty-vector cells (Wang et al., 1997). Staurosporine, H-7, calphostin C and chelerythrine were found to enhance ara-C-induced apoptosis in HL-60 and U937 cells (Grant et al, 1994). Staurosporine potentiated ara-C-related degradation of DNA to oligonucleosomal fragments in HL-60 and U937 cells, but was ineffective when given alone at these concentrations. In contrast, co-administration of H-7, calphostin C and chelerythrine, also increased the extent of DNA fragmentation observed in ara-C-treated cells, but these effects were evident only at inhibitor concentrations that were by themselves sufficient to induce DNA damage (Grant et al., 1994). Exposure of HL-60 cells to ara-C induced time- and concentration-related apoptosis. Treatment with bryostatin 1 alone failed to induce DNA damage, but promoted substantial time- and concentration-related increases in the extent of apoptosis induced by a subsequent exposure to ara-C. Concentrations of bryostatin 1 that maximally potentiated ara-C-related DNA fragmentation were associated with virtually complete down-regulation of total cellular PKC activity, whereas diglyceride and phospholipase C, which suppressed the response to ara-C, moderately increased total PKC activity (Jarvis et al., 1994b; Jarvis et al., 1998). Apoptosis of freshly isolated rat hepatocytes was induced by either the omission of fetal bovine serum in the culture medium or addition of polymyxin B or staurosporine. This effect was partially prevented by short-term treatment with TPA. After eight hours of incubation, TPA failed to counteract this action and itself produced the apoptosis of rat hepatocytes (Sanchez et al., 1992). UCN-01 enhanced cis-platin cytotoxicity and apoptosis in ovary cancer cells. This occurs regardless of p53 status, but in wild-type p53 cells the degree of sensitization seemed to be increased (Husain et al., 1997). Staurosporine and H-7 augmented TNF-mediated DNA fragmentation in HEL human embryonic lung fibroblast cells which are normally TNF resistant (Kobayashi et al., 1997). Staurosporine increased the TNF-mediated cytotoxicity in two renal cell lines (Woo et al., 1996). Staurosporine and tamoxifen induced apoptosis in glioma cell lines (Couldwell et al., 1994). Tamoxifen significantly enhanced adriamycin-induced cytotoxicity and apoptosis of hepatocellular Hep-3B MDR1 expressing cells. (Cheng et al., 1998). Safingol increased apoptosis in SK-GT-5 and MKN-74 gastric cancer cells after exposure to the alkylating agent mitomycin C. Simultaneous exposure of SK-GT-5 cells to safingol and TPA abrogated the safingol-mediated ehancement of mitomycin C-induced apoptosis. (Schwartz et al, 1995). Down-regulation of PKCα by antisense phosphorothioate oligonucleotides led to apoptosis in glioma cells (Dooley et al., 1998). The PKC inhibitor Ro 31-8220, which preferentially inhibits the PKCα, β and γ isoenzymes, induced apoptosis mainly in glioblastoma cells expressing high levels of PKCα. PKCα suppressed apoptosis in theses cells by restricting the accumulation of p53 and the expression of insulin-like growth factor-1-binding protein as well as by maintaining the retinoblastoma protein in an inactive hyperphosphorylated state (Shen and Glazer, 1998). Exposure of cells to a genotoxic stimulus that induced apoptosis, led to an inhibition of PKCζ (Berra et al, 1997). The product of the par-4 gene interacted with PKCζ and inhibited its enzymatic activity. The expression of par-4 correlated with growth inhibition and apoptosis (Diaz-Meco et al, 1996).

3.5 PKC Inhibition Prevents Apoptosis

In contrast to the reports by Wang et al. (1997) and Grant et al. (1994), (see page 28) it was also reported that H-7 and staurosporine blocked apoptosis following induction by ara-C in HL-60 cells (Kharbanda, et al, 1991). The calcium ionophore A23187 induced apoptosis in immature mouse thymocytes. H-7 inhibited the DNA fragmentation and cell death (Kizaki et al.,

1989). In murine thymocytes radiation-induced DNA fragmentation could be prevented by treatment with H-7 or staurosporine during incubation time. Incubation of irradiated cells with HA-1004, an inhibitor of cAMP-dependent protein kinase, with a minor effect on PKC did not affect the DNA fragmentation induced by irradiation. Incubation of cells with PdBu gave a dose-dependent induction of DNA fragmentation. This effect could be inhibited by staurosporine (Ojeda et al., 1992). During metanephric development the metanephric mesenchyme is programmed for apoptosis. Incubation of mesenchyme with a heterologous inducer, embryonic spinal cord prevented this DNA degradation. Phorbol esters mimicked the effects of the inducer and staurosporine prevented the effect of the inducer (Koseki et al., 1992). Downmodulation of PKC by TPA prevented apoptosis in DU-145 human prostatic carcinoma cells (Rusnak and Lazo, 1996).

3.6 PKC Isoenzymes and Apoptosis

Different PKC isoenzymes seem to be involved in the induction of apoptosis. However, the exact role of each isoenzyme is not clear at present (McConkey et al., 1994; Lucas and Sanchez-Margalet, 1995; Lavin et al., 1996; Grant et al., 1996; Deacon et al., 1997). Here the known effects of different PKC isoenzymes on apoptosis are summarized.

There are reports indicating that inhibition of PKCα activity seems to be associated with apoptosis as shown by a series of antisense oligonucleotides directed towards PKCα (see section 2.3.4). Other examples are: The PKC inhibitor Ro 31-8220, which preferentially inhibits the PKCα, β and γ isoenzymes, induced apoptosis mainly in glioblastoma cells expressing high levels of PKCα (Shen and Glazer, 1998). Tamoxifen significantly enhanced adriamycin-induced cytotoxicity and apoptosis of hepatocellular Hep-3B cells. Tamoxifen inhibited the activation of PKCα. TPA restored the membrane translocation of PKCα and abrogated the synergistic cytotoxicity of tamoxifen and adriamycin (Cheng et al., 1998). In Jurkat cells induction of apoptosis by Fas-activation was found to inhibit the ability of PKCα to phosporylate histone H1, but did not inhibit PKCε (Chen and Faller, 1999). In human erythroleukemia TF-1 cells PKCα was inactivated within 5 minutes of treatment with apoptosis-inducing levels of ionizing radiation. This postirradiation inactivation did not occur when cells were rescued from apoptosis by GM-CSF. The survival signal seemed to be mediated by PKCα but not by PKCβII or PKCε (Kelly et al., 1998). A dominant negative mutant of PKCα induced apoptosis in COS-1 cells. Expression of wild-type PKCα was able to rescue the cells from apoptosis (Whelan and Parker, 1998). bFGF and phorbol esters protected endothelial cells against radiation-induced apoptosis. bFGF mediated the translocation of PKCα form the cytosol into the membrane (Haimovitz-Friedman et al., 1994a). IL-6-starvation of T1165 and T1198 plasmacytoma cell lines led to apoptosis which was suppressed by TPA-induced PKC-activation that involved PKCα and/or PKCδ (Romanova et al., 1996). PKCα activity was found to be essential for the prevention of apoptosis in Ramos-BL Burkitt's lymphoma cells (Keenan et al. ,1999).

On the other hand there are also data showing that PKCα activity seems to be essential for apoptosis. In human prostate cancer cells, the presence of PKCα in the membrane correlated with apoptosis, the absence correlated with resistance to apoptosis (Powell et al., 1996). Safingol potentiated mito-mycin C-induced apoptosis in MKN-74 gastric cancer cells. Mitomycin treatment alone resulted in a complete loss of PKCα from the membrane and the cytosol and of PKCε from the membrane. Treatment with safingol and mitomycin C together resulted in complete restoration of PKCα and PKCε to the levels of untreated controls. TPA blocked the enhancement of mitomycin C-induced apoptosis by safingol and was accompagnied with loss of PKCα but not of PKCε. From these experiments it was concluded that restoration of PKCα but not PKCε is essential for enhancement of apoptosis (Danso et al., 1997). The steady-state population of stratified squamous epithelium is maintained by balanced cell proliferation and apoptosis. The surface epithelium of the human tonsil expressed cytoplasmic PKCα, β, δ, ε, and ζ. PKCδ and ε were most abundant in viable epithelial cells while PKCα and β were most intense in cells undergoing apoptosis (Knox et al., 1993). Treatment of parental U937 cells with TPA displayed growth arrest and differentiation into a monocyte/macrophage-like cell line, while PKCζ-overexpressing cells underwent cell death following treatment with TPA. The ability of GF109203X to inhibit TPA-induced death of PKCζ overex-pressing cells suggested that activation of a conventional isoform was neces-sary to induce apoptosis (Ways et al., 1994).

One possible explanation for the contrasting results obtained with PKCα may be that in addition to PKCα also other PKC isoenzymes may be in-volved in apoptosis. The selective depletion of PKCα in smooth muscle cells by antisense oligonucleotides against PKCα inhibited proliferation without concomitant induction of apoptosis (Leszczynski et al, 1996). However, induction of apoptosis by ceramide was associated with inactivation of PKCα and activation of PKCζ (Lee et al, 1996a). PKCα and PKCζ seem to be involved in the phosphorylation of I-κB kinase β which leads to the induc-tion of NF-κB-inducable genes (Lallena et al., 1999). Another possible expla-nation may be that PKCα is activated and down-modulated during different stages of the apoptic process. For example, it was reported that caspases regulate PKCα activity in HL-60 cells. PKCα activity was initially inhibited at one hour and subsequently activated during apoptosis. PKCβI and PKCδ were proteolytically cleaved and activated during apoptosis in these cells (Shao et al., 1997), indicating also that several PKC isoenzymes may concomitantly be involved in apoptosis. The effects of PKCα inhibition or activation may also depend on the presence and status of other factors such as raf-1 or bcl-2. PKCα phosphorylates raf-1 (Kolch et al, 1993) as well as bcl-2 (Ruvolo et al, 1998) and raf-1 also interacts with the apoptosis-preventing bcl-2 protein (Blagosklonny et al, 1997).

Expression of v-abl prevented apoptosis in a haemopoietic cell line. Calphostin C restored apoptosis. Chronic exposure to TPA did not alter survival of the cells. The PKC isoenzyme responsible for these effects was identified as PKCβII (Evans et al., 1995). Induction of apoptosis by growth to high density or by etoposide in U937 cells was associated with dephosphorylation and downmodulation of PKCβI (Whelan and Parker, 1998). However, spontaneously apoptotic U937 cells from exponentially growing cell cultures exhibited increased PKCβ and reduced PKCζ expression (Pongracz et al., 1995b). 12-deoxyphorbol-13-O-phenylacetate-20-acetate (DOPPA), an activator of PKCβI, induced apoptosis in U937 cells (Pongracz et al., 1996). In contrast to the parental HL-60 cells, PKCβ-deficient HL-525 cells were resistant to TNFα-induced apoptosis but sensitive to anti-Fas monoclonal antibody-induced apoptosis. Both cell types expressed similar levels of the TNF-receptor I, whereas the Fas receptor was detected only in HL-525 cells. Transfecting the HL-525 cells with an expression vector containing PKCβ reestablished their susceptibility to TNF-alpha-induced apoptosis (Laouar et al., 1999). Ara-C induced apoptosis and concomitantly increased membrane-bound (activated) PKCβII, but not PKCα or PKCδ. Ara-C or TPA-induced translocation of PKCβII was inhibited by edelfosine, and ara-C-induced apoptosis was stimulated by pretreatment of the cells with edelfosine. Edelfosine or antisense oligonucleotides directed toward PKCβII, but not the sense contol, enhanced ara-C-induced apoptosis. Edelfosine also inhibited stimulation of bcl-2 by TPA and enhanced the decrease in bcl-2 observed in ara-C-treated cells (Whitman et al., 1997).

PKCδ is cleaved by caspase 3 during apoptosis. This leads to the generation of an active fragment of PKCδ and growth inhibitory effects (Emoto et al., 1995; Ghayur et al., 1996). PKCδ was also activated by a caspase-dependent proteolysis during UV-induced apoptosis (Denning et al., 1998). Upon treatment with TPA, the growth of CHO cells overexpressing the PKCδ subspecies was markedly inhibited, whereas cell lines overexpressing PKCα, PKCβII, and PKCζ subspecies were not significantly affected (Watanabe et al., 1992). Bistratene A, an activator of PKCδ, induced translocation of PKCδ to the nucleus and induced growth arrest in G2/M in HL-60 cells, suggesting that activation of PKCδ can induce growth arrest and apoptosis in HL-60 cells (Griffiths et al., 1996). In rat vascular smooth muscle cells apoptosis was found to be regulated by PKCδ and ζ but not by PKCα and ε (Leszczynski et al., 1995). PKCδ was shown to be a growth and tumor repressor in rat colonic epithelial cells (Perletti et al., 1999). The tyrosine kinase and oncogene src promoted cell proliferation and DNA synthesis. Src promote tyrosine phosphorylation of PKCδ and its subsequent degradation (Blake et al., 1999).

In glucocorticoid-induced apoptosis in murine thymocytes, PKCε is selectively activated (Iwata et al., 1994). Recently also proteolytic activation of PKCε during chemotherapeutic agent-induced apoptosis by caspase 3 was reported in U937 cells (Koriyama et al., 1999). Activation of PKCε is also critical for cardiac myocyte protection from hypoxia-induced apoptosis (Gray et al., 1997). Calphostin C and TNF-α were found to induce apoptosis in U937 histiocytic lymphoma cells. Treatment with TPA prevented apoptosis induced by these compounds. A peptide derived from the V1 region of PKCε specifically blocked translocation of PKCε and blocked also the inhibitory effect on apoptosis by TPA (Mayne and Murray, 1998). These results indicate that PKCε is activated during apoptosis. However, R6 cells stably transfected with PKCε were shown to prevent cis-platin-induced apoptosis (Basu and Cline, 1995).

PKCθ is cleaved by caspase 3 during apoptosis induced by diverse agents (Datta et al, 1997). In Jurkat cells, induction of apoptosis by activation of Fas inhibited the TPA-induced translocation of PKCθ from the cytosol to the membrane (Chen and Faller, 1999). In murine thymocytes treatment with the diterpene diester, ingenol-3,-20-dibenzoate induced selective translocation of nPKCδ, ε, θ and μ from the cytosolic fraction to the particulate fraction and induced apoptosis. This induction of apoptosis was inhibited by non-isoform-selective PKC inhibitors, but not by their structural analogs with weak PKC-inhibitory activity or the selective inhibitor of cPKCs and PKCμ, Gö 6976 (Asada et al., 1998).

K562 chronic myelogenous leukemia cells are highly resistant to chemotherapeutic drugs, such as taxol, that induce cell death by apoptosis. This resistance is mediated by the chimeric tyrosine kinase oncogene bcr-abl. PKCι overexpression protected K562 cells against ocadaic acid- and taxol-induced apoptosis, whereas overexpression of PKCζ did not exhibit this resistance (Murray and Fields, 1997). Treatment of K562 cells with taxol led to sustained activation of PKCι. In contrast, bcr-abl-negative HL-60 myeloid leukemia cells, which are sensitive to taxol-induced apoptosis, did not exhibit sustained PKCι activation in response to taxol. Treatment of K562 cells with tyrphostin AG957, a selective bcr-abl inhibitor, blocked taxol-induced

PKCɩ activation and sensitized these cells to taxol-induced apoptosis, indicating that PKCɩ is a relevant downstream target of bcr-abl-mediated resistance. Furthermore, expression of constitutively active PKCɩ by adenovirus-mediated gene transfer rescued AG957-treated K562 cells from taxol-induced apoptosis (Jamieson et al., 1999). If focal adhesion kinase or the correct extracellular matrix was absent, cells entered apoptosis through a p53-dependent pathway. PKCλ/ι was required for this apoptosis. It was concluced that PKCλ/ι phosphorylates p53 and increases its stability enabling it to induce apoptosis (Ilic et al., 1998).

Exposure of cells to a genotoxic stimulus that induced apoptosis led to an inhibition of PKCζ (Berra et al, 1997). The product of the par-4 gene interacts with PKCζ and inhibits its enzymatic activity. The expression of par-4 correlated with growth inhibition and apoptosis (Diaz-Meco et al, 1996). Loss of the ζ isoenzyme triggered apoptosis (Leszczynski et al, 1995). Overexpression of PKCζ did not protect cells from taxol-induced apoptosis (Murray and Fields, 1997), indicating also that elevated levels of this isoenzyme are not necessarily correlated with resistance to apoptosis. PKCζ is cleaved by caspase 3 following induction of apoptosis by UV. The PKCζ fragment generated is enzymatically inactive. PKCɩ is not cleaved following UV treatment (Frutos et al., 1999). DNA fragmentation was associated with inhibition of PKC by H-7, chelerythrine, calphostin C in HL-60 cells promyelocytic cells. Induction of apoptotic DNA damage and cell death by activation of the sphingomyelin pathway led to increased ceramide levels (Jarvis et al., 1994a). TPA induced apoptosis and rapid ceramide generation in LNCaP prostate cancer cells. Treatment with fumonisin B1, a specific inhibitor of ceramide synthase, abrogated both ceramide production and TPA-induced apoptosis. Ceramide analogues bypassed fumonisin B1 inhibition to initiate apoptosis directly. Thus, ceramide appeared to be a neccesary signal for TPA-induced apoptosis in LNCaP prostate cancer cells (Garzotto et al., 1998). Exposure of murine B lymphoma WEHI-231 cells to chelerythrine or calphostine C induced apoptosis and increased ceramide production. This suggested an antagonistic relationship between PKC activity and ceramide in the signaling events preceding apoptosis (Chmura et al., 1996a; Chmura et al., 1996b). Ceramide a potent inducer of apoptosis (Hannun, 1998), is an activator of PKCζ (Lozano et al., 1994). An explanation for this discrepancy may be that ceramide concentrations as low as 0.5 nM activated PKCζ, whereas concentrations above 60 nM led to a downmodulation. Based on these data it was suggested that PKCζ may act as a molecular switch between mitogenic and growth inhibitory signals (Muller et al., 1995). In addition to PKC, ceramide also interacts with other targets as ceramide-activated protein kinase or ceramide-activated protein phosphatase (Mathias et al.,

1998). The exact role of ceramide in apoptosis seems not to be clear because of problems detecting ceramide levels (Kolesnick and Hannun, 1999; Hofmann and Dixit, 1999; Watts et al, 1999). Ganglioside GD3 is an important mediator of Fas-induced apoptosis in hematopoietic cells. Ceramide triggers the synthesis of ganglioside GD3 (DeMaria et al., 1997). So it is difficult to explain whether the inducer of apoptosis is ceramide or ganglioside.

3.7 Remarks and Conclusions

These contrasting results point to a great variability depending on cell type, cell environment, agent, phase of the cell cycle, and intracellular signaling pathways causing apoptosis. One reason for the conflicting results may be the use of different cell lines or even different clones of the same cell line for these investigations, and the expression of different PKC isoenzymes in these cells. For example, it has been reported that PC12 cells express PKCα, β, δ, ε, and ζ (Wooten et al., 1994). Others reported that the PC12 cells they used also expressed PKCγ (O´Driscoll et al., 1995) or PKCη (Borgatti et al., 1996). This may lead to contradicting results obtained in similar experiments employing the same cell line. The reason for controversial results may also be due to cells used from different species. For example, apoptosis was potentiated by TPA and prevented by H-7 in mouse thymocytes but not in rat thymocytes (Shaposhnikova et al., 1994). Death suppression induced by IL-6-starvation in T1165 and T1198 plasmacytoma cells by transient TPA-induced PKC activation occured when a significant number of cells were in "competent" G1 state, allowing them to pass the restriction point safely without initiating the cell death program (Romanova et al., 1996).

The microenvironment may also influence apoptosis. Exposure of isolated thymocytes to TPA plus ionomycin for 24 hours enhanced apoptosis. On the other hand, when thymocytes were cultured in intact lobes, a 24 hour TPA plus ionomycin exposure only marginally induced apoptosis. Therefore, it appears that removing thymocytes from their thymic microenvironment makes the cells more susceptible to certain stimuli, possibly by altering their physiological status. (Moore et al., 1992). Viral infection may also alter apoptosis. Epstein-Barr virus infected human Burkitt's lymphoma cells were particularly sensitive to treatment with PdBu (42% apoptosis at 72 hours), whereas its virus free counterpart displayed only 12% apoptosis (Ishii and Gobe, 1993).

Apoptosis can be induced or prevented through different pathways (TNFα-induced, Fas-induced, TRAIL-induced). In lymphokine-activated killer cells Fas-mediated cytotoxicity could be dissociated from perforin-mediated cytotoxicity by their different requirement of TPA-sensitive PKC isoforms (Ohmi et al., 1997). Apoptotic death in T cell hybridomas can be induced by glucocorticoids or the stimulation via the TCR/CD3 complex. The two apoptotic processes are mutually antagonistic (Iseki et al., 1991). Interleukin-3 and TPA rescued differentiating myeloid leukemic cells by different pathways. (Lotem et al., 1991). An ATP-dependent and an ATP-independent apoptotic pathway was found by Eguchi et al. (1999). It was reported that both, an increase in the intracellular Ca^{2+} level and an activation of PKC are essential for the TCR/CD3-mediated apoptosis. Either reduction of extracellular Ca^{2+} or addition of H-7 inhibited anti-CD3-induced but not dexamethasone-induced DNA fragmentation. The combination of ionomycin and TPA, but neither one alone nor the combination of ionomycin and cyclic nucleotide analogs, induced DNA fragmentation. On the contrary, only an increase in the intracellular Ca^{2+} level was essential for the inhibition of glucocorticoid-induced apoptosis, because ionomycin alone as well as the combination of ionomycin and TPA inhibited dexamethasone- but not anti-CD3-induced DNA fragmentation (Iseki et al., 1991). The mechanism of induction of apoptosis by staurosporine, UCN-01, and UCN-02 was clearly different from the mechanism that mediated induction of apoptosis by etoposide and dexamethasone (Harkin et al., 1998).

Staurosporine, UCN-01, and UCN-02 (a weak PKC inhibitor) induced a concentration- and time-dependent increase in apoptosis, whereas neither CGP 41251, RO 31-8220, nor GF 109203X induced apoptosis in immature rat thymocytes. In the human cell line BM13674 the specific inhibition of PKA gave rise to significantly increased levels of apoptosis at postirradiation compared to values after radiation exposure only. Calphostin C which caused 68% inhibition of PKC activity in irradiated cells, did not alter the level of radiation-induced apoptosis (Findik et al., 1995).

It was also shown that different concentrations of a drug or a protein may elicit different effects. The culture of insulin-secreting RIN m5F cells with 0.1–1 nM of staurosporine inhibited DNA synthesis but were unable to trigger apoptosis. 0.1–1 μM of staurosporine which abolished DNA synthesis almost completely, were needed to trigger apoptosis. TPA failed to inhibit this effect. (Sanchez-Margalet et al., 1993). Low concentrations of ceramide activated PKCζ, high concentrations led to a downmodulation (Muller et al., 1995). Bryostatins 1 and 2 inhibited cell proliferation at low concentrations, but showed no effect at high concentrations (Dale et al., 1989). Low levels of raf-1 activity induced proliferation whereas high levels caused growth arrest (Sewing et al., 1997). Low levels of bcl-2 were found to be antiapoptotic, high levels not (Shinoura et al., 1999).

The induction or prevention of apoptosis may depend, in addition to or independent of PKC, on calcium, pH, p53, cyclin-dependent kinases, cyclin-dependent kinase inhibitors, caspases, raf, bax, bcl-2, bcl-x, mdm2, NF-κB, pRB, STAT-factors, and many others (Fig. 2). The expression levels of all these factors are usually not determined in the cells used for experiments but they may lead to different results in similar experiments. Contrasting results may also arise from the presence or absence of wild-type or mutated p53. One hypothesis is that PKC may phosphorylate p53 and keep it in a latent inactive state (Magnelli and Chiarugi, 1997), although this has been questioned (Blattner et al., 1999). It is not clear at present whether p53 is a target for PKC phosphorylation (Livneh and Fishman, 1999).

Oncogenes or growth factors, such as c-myc, c-fos, E2F, cyclin D or E1A can induce proliferation and cellular survival but can also cause apoptosis and growth arrest. Quiescent cells may respond to these signals by proliferation whereas proliferating cells may respond by growth arrest or apoptosis. (Lavin et al., 1996; Blagosklonny, 1999). TPA is mitogenic for mature T cells and normal melanocytes. However, it causes cell cycle arrest in tumorigenic T cells and malignant melanoma cells (Burger et al., 1994; Coppock et al., 1995; Desrivieres et al., 1997).

Another reason for conflicting results may be the that the PKC mudulators available so far are not specific. Even the new generation of more specific PKC inhibitors may affect yet uncharacterized kinases (Dieter and Fitzke, 1991; Yu et al., 1998). Different cells or different clones of the same cell line express different PKC isoenzymes (Wooten et al., 1994; O´Driscoll et al., 1995; Borgatti et al., 1996) and, therefore, unspecific inhibitors display different effects. For example, dermal papilla cells underwent apoptosis in a dose-dependent manner when treated with staurosporine but not when treated with H-7 (Ferraris et al., 1997). Chelerythrine inhibited taurine uptake in the retina in a PKC-independent way (Militante and Lomardini, 1999). Induction of apoptosis by PKC modulators may be independent of PKC. For example, in human DU-145 prostatic carcinoma cells, sphingosine induced apoptosis through down-regulation of bcl-2 or bcl-X_L, independently of PKC inhibition (Shirahama et al., 1997, Sakakura et al., 1996). Harkin et al. (1998) found that inhibition of PKC alone is insufficient for induction of apoptosis in thymocytes. Data obtained with TPA are difficult to interpret, because short term treatment with the compound leads to an activation and long term treatment to a downmodulation of PKC (Rodriguez-Pena and Rozengurt, 1984). TPA also interacts with other proteins (Ahmed et al., 1993), PKD (Valverde et al., 1994) and β2-chimaerin (Caloca et al., 1997) and induces the expression of different genes (Rahmsdorf and Herrlich, 1990; Schlatterer et al., 1999). The activation or down-modulation of PKC by TPA may be vary in different cells due to the expression of different PKC isoenzymes. Usually PKC is activated and subsequently downmodulated. It was reported that during apoptosis PKC is inhibited first and subsequently activated (Shao et al., 1997).

Overexpression of one PKC isoenzyme may lead to altered expression and activity of one or more of the other PKC isoenzymes. For certain effects several PKC isoenzymes may be involved. For example, the combined effects of PKCλ, ε and ζ are essential for the transcriptional activation of c-fos by oncogenic H-ras (Kampfer et al., 1998). PKCλ and ζ participate in the ras-mediated reorganization of the F-actin cytoskeleton (Überall et al., 1999). PKCζ can control the phosphorylation and activation of PKCδ (Ziegler et al., 1999). It may be that not the levels of PKCs, but the levels of the dephosphorylated forms of PKCs are important for apoptosis (Whelan and Parker, 1998).

4 PKC and MDR

4.1 Introduction

Cells selected for resistance against a single cytostatic drug may simultaneously acquire cross-resistance to a range of other drugs (Gottesman and Pastan, 1993). Cross-resistance is related to decreased intracellular drug accumulation that is correlated with the presence of a plasma membrane 170-kilodalton glycoprotein (PGP) (Juliano and Ling, 1976). This protein is a broad specificity efflux pump encoded by multidrug resistance genes. Multidrug resistance results in resistance to major classes of anticancer drugs in clinical use, e.g., Vinca alkaloids, anthracyclins, podophyllotoxins and actinomycin D (Gottesman and Pastan, 1993).

There are many reports showing an influence of PKC on MDR. On the other hand, many publications indicated that PKC seems not play a role. Possible interferences between PKC and MDR1 or PGP are indicated in Fig. 3. Here the results showing an involvement of PKC and that arguing against it are summarized.

4.2 PKC-Activity in MDR1-Mediated Drug Resistance

Fine et al. described that the activity of PKC was 7-fold higher in multidrug resistant cells compared with the sensitive parental breast cancer cells (Fine et al., 1988). An over 6-fold increase in PKC activity in the MDR human breast cancer subline MCF-7/DOXR was confirmed when compared with the sensitive parent cell line, MCF-7/WT (Schwartz et al., 1991). Aquino et al. found that multidrug resistant HL-60/ADR cells contained 2-fold more PKC than the parental cell line (Aquino et al., 1988). In four murine UV-2237M fibrosarcoma cell lines a positive correlation between the level of PKC activity and resistance to ADR was found (O'Brian et al., 1989b). Posada et al. found that the multidrug resistant sarcoma 180A10 subline had the same quantity of PKC as the parent Sarcoma 180 cells, but the resistant cells had significantly higher intrinsic PKC activity and an altered ability to translocate the enzyme (Posada et al., 1989a). It was also shown that multidrug resistant sarcoma 180 and KB cell lines exhibited 80-90% increases in basal PKC activity (Posada et al., 1989b). Enzyme assays showed that multidrug resistant KB-V1 cells exhibited 4-fold higher PKC activity compared with the drug sensitive KB-3 subline (Chambers et al., 1990a, Chambers et al., 1990b). An elevated level of nuclear PKC was found in multidrug resistant MCF-7 human breast carcinoma cells (Lee et al., 1992). The PKC level in the MDR1-expressing cell line K562/D1-9 was higher than that in parental 562 cells (Urasaki et al., 1996).

4.3 Arguments Against a Role of PKC Activity in MDR

In HL-60 cells containing increased levels of PGP it was found that not PKC, but a novel membrane-associated protein kinase phosphorylates and regulates PGP activity (Staats et al., 1990). Staurosporine, at both subtoxic and toxic concentrations as well as at concentrations shown to be inhibitory to PKC, failed to increase drug resistance of parent and resistant MOLT-3 cells and to decrease drug resistance of MCF-7/WT and MCF-7/DOXR cells (Schwartz et al., 1991). Short-term exposure to TPA, which activated PKC 7.0-fold and 4.7-fold, respectively, in the membrane of MOLT-3 and resistant cells, resulted in small increases (rather than decreases) in resistance to adriamycin, whereas for vincristine no consistent trend was observed. Identical results were also obtained with PdBu. These results indicated that PKC activity can be decreased or increased in multidrug resistant cells. Both staurosporine inhibition and phorbol ester activation failed to produce changes in drug resistance that would be considered consistent with the resulting degree of PKC activity. PKC activity in these cells may then be unrelated to MDR (Schwartz et al., 1991). In MDR KB-A1 and KB-A10 cells 100- and 1000-fold resistant to adriamycin, respectively, PKC acitivity was similar in both resistant lines (Dolci et al., 1993). Total cytosolic PKC activity was 400% and 350%, PKCα protein expression was increased by 600% and 375% in KB-A1 and KB-C1, respectively, over the parent KB-3-1 line. A correlation between PKC and multidrug resistance was found only for cells selected in colchicine and not with those selected in other drugs (Drew et al., 1994). A combination of adriamycin and CGP 41251 reduced the number of lung metastases produced by i.v. injection of murine CT-26P or drug-resistant CT26R500 cells into nude mice. PKC activity was reduced in tumors derived from mice treated with either adriamycin or CGP 41251, but not from those derived from mice treated with the combination (Killion et al., 1995). In drug-sensitive CCRF-CEM, KB-3-1, Hela-WT, multidrug resistant CCRF–ACTD400, CCRF–VCR1000, CCRF–ADR5000, KB-8-5, KB–C1 and HeLa cells transfected with the MDR1 gene, the expression of PGP was the determinant of resistance and PKC did not contribute to the resistance (Utz et al., 1996). An increase in PKC activity in the MDR cell lines KB-A1 and KB-8-5, but not in the MDR lines C6–0.5 and C6–1V compared with their parental cell lines was observed. Cyclosporin A and S-9788 were the most active compounds on MDR reversal and were also able to inhibit PKC activity in the resistant KB as well as in all C6 cell lines. PKCα, γ and δ were increased in the resistant KB sublines. In contrast PKCα and γ were decreased in C6–1V cells, δ in the C5–0.5 line. It was concluded that an increase in PKC activity is not an absolute requirement for expression of the MDR phenotype provided that the basal level be high enough. However, it was also concluded that some modulators may act on PGP, not only through direct PGP interaction, but also through PGP phosphorylation or expression. (Hu and Robert, 1997). A human leukemia K562 mutant 100-fold resistant to the induction of apoptosis by the phosphatase inhibitor okadaic acid showed similar levels of phosphatases but lacked PKCε and overexpressed PGP indicating that MDR is not necessarily associated with increased cPKC activity (Zheng et al., 1994).

4.4 PKC Inhibition or Downmodulation in MDR

Inhibition of PKC activity by staurosporine resulted in decreased resistance to adriamycin. (Posada et al., 1989b; Sato et al., 1990; Sampson et al., 1993). H-7 completely reversed the protection against ADR cytotoxicity conferred on UV-2237M-ADRR cells by deoxycholate, providing evidence that deoxycholate exerts its protective effects by a mechanism that involves stimulation of protein phosphorylation and not merely by detergent effects on membrane permeability (O'Brian et al., 1991a). The PKC inhibitory peptide N-myristoyl-RKRTLRRL reversed adriamycin-resistance in UV-2237M-ADRR cells (O´Brian et al., 1991b). Drug accumulation assays demonstrated that in multidrug resistant KB-V1 cells TPA caused a decrease, wheras staurosporine and calphostin C caused an increase, in accumulation of [^{3}H]vinblastine. These compounds did not alter [^{3}H]vinblastine levels in drug-sensitive KB-3 cells (Chambers et al., 1992). An intrinsic resistant LoVo colon adenocarcinoma cell line showed a significant increase of PKC activity compared with the parental sensitive cell line. Preincubation with H-7 induced PKC inhibition and reversal of drug resistance (Dolfini et al., 1993). Inhibition of PKC with calphostin C, staurosporine or prolonged treatment with TPA decreased phosphorylation of PGP and impaired transport of vinblastine. Calphostin C also inhibited transport of actinomycin D, vincristine, rhodamine and azidopine in SW620 Ad300 multidrug-resistant human colon carcinoma cells. Photoaffinity labeling of PGP with azidopine was decreased by calphostin C. From these results it was concluded that dephosphorylation alters the affinity of PGP for its substrates (Bates et al., 1993). Daunorubicin resistance in differentiated blast cells was not correlated with the level of PGP expression but rather with the ability to extrude rhodamine 123. Staurosporine used at subtoxic concentrations induced a twofold to threefold enhancement of daunorubicin cytotoxicity, increased rhodamine accumulation and decreased rhodamine efflux kinetics in resistant AML cells. These effects were observed for staurosporine concentrations much lower than those required to displace the PGP-binding probe azidoprazosin, suggesting that staurosporine might act through its PKC inhibitory effect and not through PGP binding (Laredo et al., 1994). Safingol, a lysosphingolipid PKC inhibitor competitively interacts at the regulatory phorbol binding domain of PKC. Safingol treatment of sensitive MCF-7 and resistant MCF-7 DOXR cells inhibited phosphorylation of the myristoylated alanine-rich PKC substrate in both cell lines, suggesting inhibition of PKC. However, only in MCF-7 DOXR cells safingol treatment increased accumulation of [^{3}H]vinblastine and enhanced toxicity of Vinca alkaloids and anthracyclines. Drug accumulation changes in MCF-7 DOXR cells treated with safingol were accompanied by inhibtion of basal and PdBu stimulated phosphorylation of PGP. Treatment of MCF-7 DOXR cell membranes with safingol did not inhibit [^{3}H]vinblastine binding. Therefore, it was concluded that enhanced drug accumulation and sensitivity in MCF-7 DOXR cells treated with safingol are correlated with inhibition of PKC rather than competitive interference with PGP drug binding through direct interaction with PGP (Sachs et al., 1995). Sphingosine stereoisomers increased vinblastine accumulation up to 6-fold in MCF-7/ADR cells, but did not alter it in sensitive MCF-7 wild type cells. PdBu treatment of MCF7/ADR cells increased phosphorylation of PGP, and this increase was inhibited by prior treatment with sphingosine stereoisomers. Sphingosine stereoisomers did not inhibit specific binding of [^{3}H]vinblastine or [^{3}H]azidopine photoaffinity labeling, suggesting inhibition of PKC-mediated phosphorylation (Sachs et al., 1996). Three N-myristoylated peptides corresponding to the pseudosubstrate region of PKCα restored intracellular accumulation of chemotherapeutic drugs in MCF-7/MDR cells in association with inhibition of the phosphorylation of PKCα substrates. A fourth peptide did not affect drug accumulation and failed to inhibit the phosphorylation of the PKCα substrates. An effective peptide did not bind to PGP as shown by its inability to inhibit [^{3}H]azidopine photoaffinity labeling (Gupta et al., 1996). Nontoxic concentrations of CGP 41251 significantly enhanced the cytotoxic properties of adriamycin, actinomycin D, vinblastine, vincristine but not those of 5-FU (5-fluorouracil). CGP 41251 increased intracellular adriamycin but did not cause significant differences in PGP expression. Pretreatment of MCF-7/ADR cells with TPA reduced CGP 41251-mediated intracellular accumulation of adriamycin and decreased the level of PGP phosphorylation but did not compete with azidopine. The conclusion of this investigations was that CGP 41251 reverses the MDR phenotype by modulating the phosphorylation of PGP or other PKC substrates critical to the maintenance of the MDR phenotype (Beltran et al., 1997).

4.5 Arguments Against a Role of PKC Inhibition in MDR

The isoquinoline PKC-inhibitors H-7, H-8 and H-9 did not reverse resistance to vinblastine in multidrug resistant K562–ADR and P388/ADR cells. It was found that reversal of MDR by isoquinoline derivatives did not correlate with the reversal of resistance and that they reverse resistance due to the suppression of drug binding to PGP (Wakusawa et al., 1992). In another study, H-8, H-9 and H-86 reversed the resistance of P388/ADR cells. These compounds dose-dependently inhibited photaffinity labeling, indicating a direct interaction with PGP (Nakamura et al., 1993). Several staurosporine derivatives enhanced accumulation of vinblastine in P388/ADR cells in a dose-dependent manner. The potency of these compounds correlated with inhibition of [^{3}H]azidopine photolabeling, but was not correlated with their inhibitory activity on protein kinases (Miyamoto et al., 1993; Wakusawa et al., 1993). The PKC inhibitory staurosporine-derivative NA-382 completely reversed the vinblastine resistance of P388/ADR cells without effect on the parental P388 cells. Photolabeling experiments with [^{3}H]azidopine suggested that NA-382 reverses MDR by direct inhibition of the drug efflux system of PGP (Miyamoto et al., 1992). Calphostin C inhibits PKC only if it is activated by light. Calphostin C (without prior exposure to light) increased the accumulation of daunorubicin in P388/ADR cells in a concentration-dependent manner. This effect was not observed in HL-60/AR cells expressing MRP. Calphostin C increased the uptake and decreased the efflux of rhodamine 123 in P388/ADR cells but had no such effect in drug-sensitive P388 cells. These data suggested that calphostin C may reverse drug resistance independently of its effect on PKC activity (Gupta et al., 1994). In mouse lymphoma HU-1 cells transfected with MDR1 cDNA, among the tested indole carbazole (K-252a) family of protein kinase inhibitors, only KT-5720 could overcome MDR. Since other protein kinase A, C and G modulators did not reverse MDR, it was concluded that the chemosensitising activity of KT-5720 on these cells is apparently independent of its kinase inhibitory effects (Galski et al., 1995). The PKC inhibitors staurosporine, CGP 41251, UCN-01, Ro-31-8820 and GF 109203X were compared in terms of their MDR-reversing properties, their susceptibility towards PGP-mediated efflux from MCF-7/ADR cells, their binding to PGP and their ability to inhibit PKC. The results suggested that these compounds affect PGP directly and not via inhibition of PKC (Budworth et al., 1996). The same conclusions were drawn after comparisons of MDR reversal, azidopine-competition and PKC inhibition of GF 109203X, dexniguldipine and dexverapamil (Gekeler et al., 1996). In HL-60 wild-type cells, HL-60 cells expressing MDR1 and HL-60 cells expressing MRP, araC-induced apoptosis can be stimulated by PKC- and PTK-inhibitors, suggesting that this process is mediated at least partially, also by PKC and PTK-independent mechanisms (Hunakova et al., 1996). The protein kinase C inhibitor NPC 15437 led to nuclear accumulation of daunorubicin and decreased LD_{50} for vincristine in multidrug resistant CH(9)C5 and MCF-7/ADR cells. Treatment with TPA partially reversed the effect of NPC 15437, suggesting that NPC 15437 was exerting an effect through protein kinase C. However, NPC 15437 inhibited the binding of [^{3}H]azidopine to PGP (Sha et al., 1996). Ilmofosine, inhibiting PKC by interaction with the regulatory region did not compete with azidopine and did not reverse multidrug resistance (Hofmann et al., 1997). Dexniguldipine-HCl reversed resistance by direct interaction with PGP (Hofmann et al., 1995). The PKC-inhibitory N-myristoylated PKCα psuedosubstrate peptide potently and selectively induced the uptake of cytotoxic drug in colon cancer cells devoid of PGP expression (Bergman et al., 1997). The staurosporine-derivatives CGP 41251 and CGP 42700 reversed multidrug resistance to a similar extent, although CGP 42700 did not inhibit PKC (Utz et al., 1998). In LoVo cells long term inhibition of PKCα and βI by Gö6976 (Martiny-Baron et al., 1993) led to increased survival following treatment with adriamycin (La Porta et al., 1998). There are no reports to date that UCN-01 also reverses MDR. The bisindolylmaleimide protein kinase inhibitor Ro 32–2241 was found to reverse MDR1-mediated MDR by acting directly on PGP rather than, or in addition to, an effect on PKC (Merrit et al., 1999).

4.6 Phorbol Ester and MDR Modulation

As mentioned in chapter 2.3.1, in many cases it is difficult to distinguish whether the effects of phorbol esters are due to activation or inactivation of

PKC. Despite this disadvantage many experiments in MDR1-mediated resistance have been performed with TPA or PdBu. Experiments with TPA suggested that the PKC effect is linked to drug sensitivity, since activation of the enzyme by short TPA exposure enhanced adriamycin's cytotoxicity as well as its ability to provoke DNA damage. Likewise, down-regulation of PKC by extended TPA exposure partially protected the cells from adriamycin-induced cytotoxicity as well as from DNA damage. Thus, the ability of cells to be injured by adriamycin appeared to be correlated with the activity of PKC (Posada et al., 1989b). TPA-induced downregulation of PKC activity was less in MDR UV-2237M-ADRR cells (due to reduced rates of PKC degradation) compared to the parental UV-2237M cells (Ward and O'Brian, 1991). The MDR1-expressing cell line KM12L4a exhibited significantly reduced sensitivity to adriamycin, vincristine and vinblastine, but not to 5-FU, following 96-hour incubation with 15 nM PdBu. Because 15 nM PdBu did not downmodulate PKC, it was concluded that this effect was due to activation of PKC. Treatment of the cells with diacylglycerol reduced [^{14}C]adriamycin and [^{3}H]vincristine accumulation significantly, which were completely reversed by H-7 (Dong et al., 1991). A 2.5 hours exposure to PdBu activated PKC and induced a 4-fold transient MDR (Morgan et al., 1991). The PKC activators TPA and DAG increased the activity as well as the levels of PGP in several cell lines derived from leukemias and solid tumors. The increase was observed at the mRNA and protein level and was suppressed by staurosporine (Chaudhary and Roninson, 1992). In K562 cells TPA was found to increase the transcription of the MDR1 gene through activation of the MDR1 promoter (McCoy et al., 1995; McCoy et al., 1999). Cells exposed to 100 nM of TPA for one hour were approximately 3-fold more resistant to adriamycin than cells exposed to adriamycine alone. The PKC inhibitor H-7 completely blocked the TPA-induced effect, but did not reverse the MDR phenotype. TPA-treated cells showed significantly higher levels of expression of PGP when compared to those from control cells (Ahn et al., 1996). Drug accumulation assays revealed that TPA treatment of KB-V1 cells significantly reduced [^{3}H]vinblastine accumulation induced by verapamil. From these results it was concluded that PKC-mediated phosphorylation stimulates the drug transport activity of PGP (Chambers et al., 1990a, Chambers et al., 1990b). Pretreatment of MCF-7/ADR cells with TPA reduced the CGP 41251 mediated intracellular accumulation of [^{14}C]adriamycin. At concentrations that induced drug uptake, CGP 41251 significantly decreased the level of PGP phosphorylation in the cells but did not compete with [^{3}H]azidopine for photoaffinity labeling of PGP. From these data it was concluded that CGP 41251 reverses the MDR phenotype by modulating the phosphorylation of PGP and/or other PKC substrates critical to the maintenance of the MDR phenotype (Beltran et al., 1997). The results presented in this chapter indicate that phorbol esters or modulation of PKC by phorbol esters alter MDR1-mediated resistance. However, there are also results indicating that phorbol esters seem not to be involved in the modulation of resistance.

4.7 Results Indicating no Influence of Phorbol Esters in MDR Modulation

The human HL-60 R1B6 subclone resistant to PdBu exhibited PKC activity following twenty four to thirty six hours after removal of PDBu from the medium. Despite differences in PKC activity there was no difference in the cellular accumulation of daunomycin or in the sensitivity to the toxic effects of adriamycin between the R1B6 cells and the parental HL-60 cell line (Hait and DeRosa, 1991). Rat1 cells treated with TPA showed neither increased MDR1 mRNA expression nor stimulation of PGP function (Kopnin et al., 1995). Phorbol esters sensitized MCF-7/ADR cells to PGP substrate drugs, however, there was no correlation with activation of PKC (Smith and Zilfou, 1995). In Rat1 fibroblasts, rat IAR2 epithelial and rat McA RH 7777, MDCK, K562 and LIM1215 (human colon carcinoma) cells, TPA showed effects in opposite directions (Stromskaya et al., 1995). TPA reduced daunomycin accumulation in both drug-sensitive KB-3–1 and multidrug resistant KB-C1 cells. TPA had no effect on daunomycin efflux and did not induce PGP expression. The results suggested that PKC may regulate drug resistance by reducing drug influx in a PGP-independent manner. This may represent a mechanism of drug resistance independent of, or in addition to, PGP-mediated drug efflux (Drew et al., 1996). TPA caused a decrease in the cellular accumulation of daunorubicin and etoposide, both in PGP-overexpressing and wild-type cells. Since treatment of cells with staurosporine reversed this effect and the non-PKC-stimulating 4α-phorbol-12,13-didecanoate did not result in a decreased daunorubicin accumulation it was concluded that this effect is the result of kinase activity. Accumulation of the PGP substrate calcein-acetoxymethyl-ester was not influenced by TPA in wild-type cells. Activation of PKC with TPA or inhibition of protein phosphatases 1 and 2A by okadaic acid did not affect the accumulation of calcein-acetoxymethyl-ester in the MDR or wild-type cells. Staurosporine increased the calcein acetoxymethyl ester accumulation only in the MDR cells. Neither stimulating PKC with TPA nor inhibiting phosphatases with okadaic acid led to a decreased inhibition of PGP by staurosporine, indicating that staurosporine inhibits PGP directly. From these experiments it was concluded that PKC and phosphatase activity do not regulate the drug transport activity of PGP and that TPA-induced PKC activity decreases cellular drug accumulation in a PGP-independent manner (Wielinga et al., 1997). TPA significantly reduced the uptake of adriamycin and vincristine in human colon cancer cells devoid of PGP activity (Bergman et al., 1997). Bryostatin, similar to TPA, modulates PKC activity (Blumberg, 1991). Bryostatin 1 affected PGP phosphorylation, but did not reverse multidrug resistance multidrug resistance (Scala et al., 1995). It was shown that bryostatin is a potent modulator of multidrug resistance in two cell lines overexpressing a mutant MDR1-encoded PGP (valine instead of glycine in position 185), namely KB-C1 cells and HeLa cells transfected with a MDR1-V185 construct. This reversal was independent of PKC. Bryostatin 1 was not able to reverse the resistance of cells overexpressing the wild type form (glycine in position 185) of PGP, namely CCRF-ADR5000 cells and HeLa cells transfected with a MDR1-G185 construct. Treatment of mutant PGP expressing HeLa-MDR1-V185 cells with bryostatin 1 was accompanied by an increase of the intracellular accumulation of rhodamine 123, whereas no such effect could be observed in HeLa-MDR1-G185 cells (expressing wild-type PGP). HeLa-MDR1-V185 cells expressed the PKC isoforms α, δ and ζ. Downmodulation of PKCα and δ by TPA did not affect the drug accumulation by bryostatin. In HeLa-MDR1-V185 cells, short-term exposure to bryostatin 1 which led to a PKC activation was as efficient in modulating the pumping activity of PGP as long-term exposure leading to PKC depletion. These results suggested that PGP is another target of bryostatin in addition to and independent of PKC (Spitaler et al., 1998). Neither TPA, dioctylglycerol, nor staurosporine of H-7 altered intracellular drug accumulation in renal proximal tubule cells (Miller et al., 1998).

4.8 Influence of PGP Phosphorylation on MDR

The multidrug resistant sublines from both mouse sarcoma 180 and human KB lines exhibited 80-90% increases in basal PKC activity due to higher levels of PKCα protein. Inhibition of endogenous PKC activity by staurosporine resulted in decreased resistance to adriamycin. Phosphorylation of MDR cell membrane vesicles by purified PKC resulted in a level of phosphorylation of P-glycoprotein that was greater than the endogenous phosphorylation level. (Posada et al., 1989b). In HL-60 cells isolated for resistance to vincristine, staurosporine induced a major inhibition in the phosphorylation of PGP. Further studies showed that under the same conditions in which staurosporine inhibited PGP phosphorylation there has been a concomitant increase in cellular drug accumulation and a major inhibition in drug efflux (Ma et al., 1991). Staurosporine inhibited the effects of TPA on the phosphorylation of PGP and on the accumulation of vinblastine (Aftab et al., 1994). In isolated membranes, phosphorylation of PGP by purified PKC was rapid, and time-dependent dephosphorylation was inhibited by okadaic acid, an inhibitor of type 1 and type 2A protein phosphatases. In [^{32}P]-labeled intact KB-V1 cells, PGP phosphorylation was stimulated by both TPA and okadaic acid. Two-dimensional thin layer tryptic phosphopeptide maps indicated that the sites of phosphorylation were similar in control, TPA-treated, and okadaic acid-treated cells and that they corresponded to those phosphorylated by PKC in vitro. Staurosporine, calphostin C and edelfosine, inhibited PGP phosphorylation in vitro and in intact cells. Drug accumulation assays demonstrated that in KB-V1 cells TPA caused a decrease, whereas staurosporine and calphostin C caused an increase, in accumulation of [^{3}H]vinblastine. These compounds did not significantly alter [^{3}H]vinblastine levels in drug-sensitive KB-3 cells. These results suggested that PKC is chiefly responsible for PGP phosphorylation in KB-V1 cells, that membrane-associated protein phosphatases 1 and 2A are active in dephosphorylation of PGP, and that phosphorylation of PGP may be an important mechanism for modulation of drug-pumping activity (Chambers et al., 1992). Membrane vesicles from multidrug resistant KB-V1 cells were incubated with purified PKC and [γ-^{32}P]ATP. PGP was purified and serines 661, 667, 671, 675 and 683 were found to be phosphorylated (Chambers et al., 1993). Okadaic acid and calyculin A, inhibitors of protein phosphatases 1 and 2A caused mitotic arrest of HL-60, HL-60/ADR and K562 cells by chromatid scattering/overcondensation and abnormal mitotic spindles. Protein phosphorylation experiments in intact cells revealed that in multidrug resistant HL-60/ADR cells overall phosphorylation of nuclear proteins was higher than that in drug -sensitive HL-60 cells (Sakurada et al., 1992). TPA increased the phosphorylation of PGP 6-fold and selectively decreased the accumulation of vinblastine in MCF-7/ADR cells. The actions of TPA did not require new synthesis of PGP, and had similar effects in MCF-7/BC-19 cells transfected with a cDNA for PGP. Transfection of MDR1 expressing MCF-7 cells with an expression vector containing PKCα antisense reduced PKCα levels and decreased total PKC activity by 75%. This was accompanied by reduced phosphorylation of PGP, a 2-fold increase in drug retention, and a 3-fold increase in adriamycin cytotoxicity (Ahmad and Glazer, 1993). Increased resistance in PKCα-transfected BC-19 cells was associated with enhanced PKC activity, phosphorylation of PGP and decreased drug accumulation (Yu et al., 1991). PGP and PKC were coimmunoprecipitated from the multidrug resistant cell lines MCF-7/ADR and KB-V-1, using antibodies to either protein. PKCα, β, γ, ε and θ, but not δ, μ, ζ and λ were found to coimmunoprecipitate with PGP. These studies indicated that

PGP closely interacts with PKC and serves as a substrate. The association between the two proteins was enhanced by TPA (Yang et al., 1996).

4.9 Arguments Against an Involvement of PGP Phosphorylation

Transfection of BC-19 cells with PKCγ led to a 19-fold increase in PKC activity, but did not confer increased resistance to adriamycin (Ahmad et al., 1992). Ser-667, Ser-671 and Ser-683 of PGP were not only phosphorylated by PKC (Chambers et al., 1993), but also by PKA in a cAMP-dependent manner (Chambers et al., 1994). Several phorbol esters sensitized MCF-7/ADR cells to PGP substrate drugs. However, there was no correlation with activation of PKC and the sensitization was not antagonized by staurosporine. Mezerein, K-252a and H-89 sensitized MCF-7/ADR cells, increased intracellular accumulation of [^{3}H]vinblastine and antagonized photolabeling of PGP by [^{3}H]azidopine. Therefore, phosphorylation did not appear to play a significant role in regulation PGP activity in these cells (Smith and Zilfou, 1995). In P388 cells treated with increasing concentrations of bryostatin 1 the phosphorylation of PKC-specific substrate was decreased up to 94%, compared with the parental cell line. Similar decreases were observed for PKC isoenzyme expression. There was no significant degree of cross-resistance to daunorubicin in the bryosatin 1-resistant cell lines (Prendiville et al., 1994). Bryostatin 1 was also found to affect PGP phosphorylation, but did not reverse MDR (Scala et al., 1995). A matched pair of mammalian cell lines was generated expressing wild-type PGP or a non-phosphorylatable mutant protein. Mutation of the phosphorylation sites did not alter PGP expression or its subcellular localization. The transport properties of the mutant non-phosphorylatable and wild-type proteins were indistinguishable (Goodfellow et al., 1996; Germann et al., 1996). A MDR variant of the human Saos-2 osteosarcoma cell line exhibited increased levels of PGP at the plasma membrane and in the nucleus. Cellular and nuclear PKC were not modified with respect to sensitive cells indicating that resistance is not dependent on PKC (Zini et al., 1997). PGP seems to be associated with volume-activated chloride currents. It seems to be a PKC-mediated channel regulator. Therefore, phosphorylation of PGP might be important only for regulation of chloride currents (Hardy et al., 1995) and not for regulation of drug efflux.

4.10 Effects of PKC on the Levels of MDR1-mRNA and PGP

Drug-mediated MDR1 induction was blocked by nonspecific PKC inhibitors that are active against PKC, but not by a protein kinase inhibitor ineffective against PKC (Chaudhary and Roninson, 1993). In KB cells transfected with the MDR1 gene promoter and a chloramphenicol acetyltransferase reporter gene, H-7 inhibited the activation of the promoter by ethylmethane sulfonate, 5-FU or UV irrradiation (Uchiumi et al., 1993). In 18 primary renal cell carcinoma cell lines a high PKC expression significantly correlated with both resistance to adriamycin and high PGP expression (Efferth and Volm, 1992). Between the expression of PKC in 83 untreated solid human non-small cell lung carcinomas and the resistance to adriamycin a statistically significant correlation was found (Volm and Pommerenke, 1995). Davies et al. concluded from their experiments that some regulation of PGP expression at the post-translational level and a coregulation of PGP expression by PKC occurred (Davies et al., 1996). The PKC level in the MDR1-expressing cell line K562/D1-9 was higher than that in parental 562 cells (Urasaki et al., 1996). Staurosporine prevented araC-induced MDR1 overexpression (Walter et al., 1997). MCF-7/ADR cells cultivated in absence of drugs lost their resistance gradually with time, so that by week 24 they had almost completely regained the drug sensitivity seen in wild-typ MCF-7 cells. PGP levels measured by Western blot mirrored the change in adriamycin sensitivity. PGP was not detectable anymore at week 24. MCF-7/ADR cells expressed more $PKC\alpha$ and $PKC\theta$ than wild-type cells and possessed a different cellular localiziation of $PKC\varepsilon$. The expression and distribution pattern of these PKCs reverted back to that seen in wild-type cells by week 24. The results suggested that MCF-7/ADR cells lost MDR1 gene expression and PKC activity in a co-ordinate fashion, consistent with the existence of a mechanistic link between MDR1 and certain PKC isoenzymes (Budworth et al., 1997).

4.11 Arguments against a Role of PKC on MDR1-mRNA and PGP Levels

The staurosporine-derivative CGP 41251 has been shown to exert a high degree of selectivity for inhibition of PKC and PKC-mediated cellular events. The IC_{50}-values of enzyme inhibtion in vitro for PKC, protein kinase A, phosphorylase kinase, S6 kinase and tyrosine-specific protein kinase of the epidermal growth factor receptor are 50 nM, 2.4 µM, 48 nM, 5 µM and 3 µM, respectively (Meyer et al., 1989). CGP 41251 reversed multidrug resistance in CCRF-VCR1000 and KB-8511 cells. Preincubation with the compound for 12 or 24 hours did not alter MDR1-mRNA levels (Utz et al., 1994). The PKC inhibitors CGP 41251 (Utz et al., 1994), dexniguldipine-HCl (Hofmann et al., 1995) and ilmofosine (Hofmann et al., 1997) did not alter the MDR1 expression. CGP 41251 has been found to increase intracellular adriamycin but did not cause significant differences in PGP expression (Beltran et al., 1997). Inhibition of $PKC\alpha$ and βI by Gö6976 even increased the expression of PGP and the resistance to adriamycin (La Porta et al., 1998).

4.12   The Role of PKC Isoenzymes on MDR1-Mediated MDR

PKCα was modestly increased (approximately 65%) in the multidrug resistant KB-V1 cell line compared with the sensitive KB-3 cells (Cloud-Heflin et al., 1996). In MCF-7-MDR cells the MDR phenotype was associated with a 10-fold increase in PKCα activity and a 10-fold decrease in calcium-independent PKC activity due to decreased expression of PKCδ and PKCε. Phosphorylation of PGP was increased more than 20-fold in the MCF-7-MDR cell line and its phosphorylation corresponded to the increases in PGP pump function underscoring the role of PKCα (Blobe et al., 1993). MCF-7/ADR cells expressed more PKCα and PKCθ than wild-type cells and possessed a different cellular localiziation of PKCε (Budworth et al., 1997). Dolfini et al. (1993) concluded from experiments with human LoVo colon adenocarcinoma cells a contribution of PKCα to the resistance. Transfection of the MDR1 gene in sensitive breast carcinoma cells did not contribute to a high degree of resistance, whereas cotransfection of the MDR1 gene with the PKCα gene induced a high degree of resistance to anticancer drugs (Yu et al., 1991). Expression of antisense cDNA for PKCα reduced drug resistance (Ahmad and Glazer, 1993). Activation of PKC by PdBu or thymeleatoxin induced resistance to multiple anticancer drugs in the metastatic human colon cancer cell line KM12L4a cells. The induction of resistance by thymeleatoxin was associated with a reduction in cytotoxic drug accumulation in KM12L4a cells. These cells contain only PKCα from the thymeleatoxin activating isoforms. Thus, it was concluded that that activation of PKCα is sufficient for the induction of resistance observed in KM12L4a cells (Gravitt et al., 1994). Basal PKC activities and immunoreactivities of PKCα and PKCζ were higher in a multidrug resistant compared to three multidrug sensitive human glioma cell lines. PKCβ, γ and ε were not detected in these cell lines. Treatment of multidrug resistant glioma cells with 100 nM TPA for 2 hours resulted in activation of PKCα, but not of PKCζ, with concomitant decrease in vincristine accumulation and increase in PGP phosphorylation. The treatment of multidrug resistant cells with 100 nM calphostin C for 2 hours decreased immunoreactive PKCζ, but not PKCα, inducing an increase in vincristine accumulation with a concomitant decrease in PGP phosphorylation. There was no significant change in vincristine accumulation in sensitive cells treated with TPA or calphostin C. It was concluced that PKCα and ζ are involved in PGP phosphorylation and vincristine efflux (Matsumoto et al., 1995). On the other hand, the expression of PKCα did not correspond to the expression of MDR1 or to the drug-sensitivity of sensitive and multidrug resistant CCRF and KB cell linies (Gekeler et al., 1996).

Rat embryo fibroblasts transfected with PKCβI displayed elevated PKC activity. These cells exhibited significant resistance to adriamycin, actinomycin D, vinblastine and vincristine but not to 5-FU. Intracellular accumulation of the MDR-drugs was decreased, but this was not associated with an altered level of PGP expression (Fan et al., 1992). P388/ADR and drug sensitive P388 cells were permeabilized and incubated with rabbit anti-PKCα or anti-PKCβ antibodies. An anti-PKCβ antibody partially corrected the drug accumulation defect and completely reversed resistance to daunorubicin. An anti-PKCα antibody had no effect (Gollapudi et al., 1995). MDR OAW-tax cells exhibited higher levels of PKCβII compared with parental OAW-42 and OAW-dox cells (Masanek et al., 1997). In relapsed state acute myelogenous leukemias highly significant positive correlations between the expression of MDR1/PKCη were found (Beck et al., 1996; Beck et al., 1998; Spitaler et al., 1999).

4.13 Remarks and Conclusions

One of the possible explanations of the contradicting results is the fact that most of the results were obtained with tumor cells. These cells have very unstable genoms and alterations in gene expressions are common. Overexpression of MDR1 and one of the eleven PKC isoenzymes may be accidential. For example, it was found that a correlation between PKC and multidrug resistance exists only in cells selected in colchicine and not with those selected in other drugs (Drew et al., 1994). In LoVo cell clones exhibiting MDR1-mediated resistance PKCα was increased and PKCε was decreased. However, a similar pattern of PKC expression was also observed in a LoVo clone exhibiting MRP-mediated resistance (Dolfini et al., 1997). Antisense oligonucleotides directed to PKC isoenzymes reduce MDR. However, used in non-MDR1 expressing cells they also induce cell death (Ahmad et al., 1994; Yazaki et al., 1996; Dean et al., 1996; Leszczynski et al., 1996; Dooley et al., 1998).

Several PKC effects were obtained from conclusions which are not substantiated by experiments. If the photoaffinity labeling with azidopine is decreased by a PKC inhibitor which reverses MDR, it may be concluded that the inhibitor competes with azidopine by binding to PGP. However, it can also be concluded that the PKC inhibitor reduces the phosphorylation of PGP and that, due to altered affinity of unphosphorylated PGP to drugs, less azidopine binds to PGP. If a PKC inhibitor reverses MDR, it may be concluded that the reversal is due to inhibition of PKC and reduced PGP phosphorylation. However, a PKC inhibitor might also reduce the drug efflux by direct interaction with PGP. There are several possibilities how a PKC inhibi-

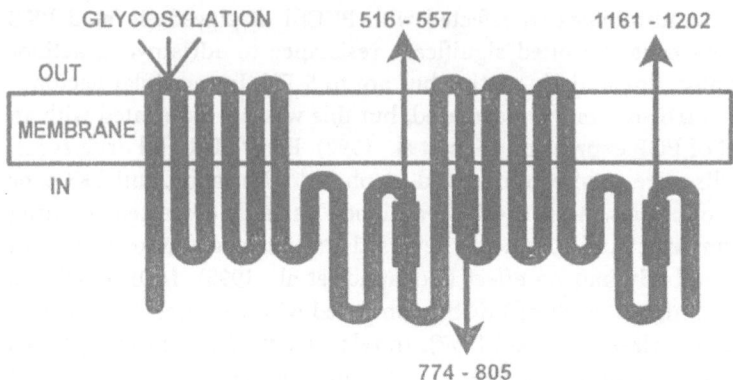

Fig. 4. Schematic representation of potential ATP binding sites in human PGP

tor might reduce drug efflux by direct interaction with PGP. Such an inhibitor might block the drug efflux by interaction with the drug binding site. In theory it might block a potential phosphorylation site which is responsible for activation of the efflux. Another possibility is that the PKC inhibitor interacts with the ATP binding site of PGP. Many PKC inhibitors such as staurosporine, CGP 41251 or UCN-01, are believed to compete with ATP at the ATP binding site of PKC. The drug efflux mediated by PGP is also dependent on ATP. PGP contains two ATP binding sites (amino acids 516–557 and 1161–1202; Gottesman and Pastan, 1993) to which binding of ATP is essential for drug transport (Fig. 4). As shown in the sequence comparison in Table 1 (below), several amino acids in the ATP binding sites of PKC and PGP are identical. So it is conceivable that a competition of a PKC inhibitor for the ATP binding site of PGP reduces drug efflux. A sequence analysis shows that in addition to these two ATP binding sites in PGP, there is a third sequence which is related to the ATP binding sites of PKCs (amino acid sequence 775–805, Table 1). Fig. 4 shows the location of these sequences. In theory, a PKC inhibitor could also bind to this site (775–805) and affect the drug efflux. So PKC inhibitors might interfere with MDR in different ways which are not known at present. The lack of this knowledge may lead to different conclusions as described above.

The effects of PKC on PGP may be time-dependent. For example, it has been reported that long term inhibition of PKC enhanced PGP expression and MDR (La Porta et al., 1998). If PKC activity is elevated in MDR1 overexpressing cells and this leads to increased resistance compared with sensitive cells, it could be speculated that PKC activity might increase the resistance by mechanisms independent of MDR1, for example, by reduction of apoptosis (due to expression of bcl-2). Castro et al. (1999) reported that PKC

Table 1. Sequence comparisons of potential ATP binding sites in PKC isoenzymes and PGP (Amino acids occurring in both, PKC and PGP, in a similar position are printed in bold)

PKC α	LGKG--S-FGKVM--LADR-KG-TEELY--AIKIL-KK-----D
PKC β	LGKG--S-FGKVM--LSER-KG-TDELY--AVKIL-KK-----D
PKC γ	LGKG--S-FGKVM--LAER-RG-SDELY--AIKIL-KK-----D
PKC δ	LGKG--S-FGKVL--LGEL-KG-RGE-YS-AIKAL-KK-----D
PKC ε	LGKG--S-FGKVM--LAEL-KG-KDEVY--AVKVL-KK-----D
PKC θ	LGKG--S-FGKVF--LAEF-KK-TNQFF--AIKAL-KK-----D
PKC L	LGKG--S-FGKVM--LARV-KE-TGDLY--AVKVL-KK-----D
PKC ι	IGRG--S-YAKVL--LVRL-KK-TDRIY--AMKV-VKK-----E
PKC ζ	IGRG--S-YAKVL--LVRL-KK-NDQIY--AMKV-VKK-----E
PGP 775	L-QGF-T-FGKAGEILT---KRLR---YMVFRSML-RQ-----D 805
PGP 516	LPHKFDTLVGERGAQLSGGQKQ-R--IA-IAR-ALVRNPKILLLD 557
PGP 1161	LPNKYSTKVGDKGTQLSGGQKQ-R--IA-IAR-ALVRQPHILLLD 1202

inhibitors, in addition to direct interaction with PGP, may reverse MDR by a pathway that involves inhibition of PKC, but is independent of PGP phosphorylation. This might be due to PKC-mediated phosphorylation of one or more proteins that modulate PGP (Castro et al., 1999). Another possible reason for contradictory results may be that increased intracellular drug concentrations may not represent the drug concentrations in the nucleus where the DNA binding occurs.

Overexpression of MDR1 may influence experimental results through additional effects of PGP, independent of drug efflux. In addition to the development of MDR, exposure of tumor cells to antimitotic agents produces further cellular changes, as activation of the MAPK pathway and acceleration of the cell cycle machinery, that may contribute to the failure of chemotherapy (Emanuel et al., 1999). It has been reported that overexpression of PGP delays the apoptotic cascade in CHO fibroblasts (Robinson et al, 1997), leads to altered expression of genes regulating apoptosis in human myeloid leukemia cells (Kim et al., 1997) and protects CCRF-CEM and K562 cells from caspase-dependent apoptosis (Smyth et al., 1998; Johnstone et al., 1999). On the other hand, it has been reported that the somatostatin analogue TT-232 induced apoptosis in multidrug resistant and sensitive hepatocellular carcinoma cell lines to a similar level, indicating that the machinery involved in apoptosis is functional in all these cells to a similar extent (Diaconu et al., 1999). 2-deoxy-D-glucose preferentially induced apoptosis in MDR1-expressing cells (Bell et al., 1998). MDR cells typically have elevated intracellular pH and decreased plasma membrane potential and this is apparently due to PGP overexpression and not exposure to chemotherapeutic drugs. One model for PGP´s function (partitioning model) suggests that alterations in pH and/or membrane potential that accompany the overexpression of PGP, indirectly affect partitioning of chemotherapeutic drugs, but PGP does not directly pump them. It has also been shown that PGP expression leads to significant resistance to complement-mediated cytotoxicity induced by the elevation of the intracellular pH and that PGP expression alone cannot account for MDR1-mediated resistance (Weisburg et al., 1999). Multidrug resistant MCF-7 cells exhibited marked accumulation of glucosylceramide compared with parental cells. Reversal of resistance by tamoxifen, verapamil and cyclosporin A, decreased the levels of glucosylceramide. The glucosylceramide synthesis inhibitors 1-phenyl-2-palmitoylamino-3-morpholino-1-propanol or tamoxifen sensitized resistant MCF-7 cells, indicating that the presence or absence of other targets in addition to PGP may influence reversal of resistance (Lavie et al., 1997).

In K562 cells TPA activated the MDR1 promoter and led to increased MDR1 mRNA levels. This activation was mediated by the zinc finger transcription factor EGR1. This activation by TPA was inhibited by the Wilms´ Tumor suppressor WT1 (McCoy et al., 1995; McCoy et al., 1999). Controversial response to TPA may depend on the presence or absence of EGR1 or WT1 in the cell lines investigated.

5 PKC Modulation in Drug Resistance not Mediated by MDR1

5.1 PKC Inhibition Enhances the Sensitivity to Antitumor Drugs

Down-modulation of PKC by long-term treatment with TPA, PKC inhibitors, such as quercetin, tamoxifen, staurosporine and the ether lipid ilmofosine enhanced the antiproliferative activity of antitumor drugs in vitro and in human tumor xenografts in nude mice (Hofmann et al, 1988; Hofmann et al., 1989; Hofmann et al., 1990). Noseda et al., (1988) found additivity in a combination of ilmofosine with cis-platin. Quercetin synergistically inhibited the growth of HL-60 cells if combined with ara-C (Teofili et al., 1992). However, Freund et al. (1998) questioned these results using the same cell line. Enhancement of cytotoxic drugs by quercetin may be due to the induction of transforming growth factor β1 or inhibition of tyrosine kinases (Larocca et al., 1995; Ferry et al., 1996). A combination of quercetin with carboplatin has been investigated in a clinical phase I trial (Ferry et al., 1996; Fyfe et al., 1996). The PKC inhibitor SPC-100270 enhanced the antiproliferative activity of adriamycin or cis-platin on murine isografts and human melanoma, lung adenocarcinoma or protstate tumor xenografts (Adams et al., 1993). The compound did not potentiate the cytotoxicity of either adriamycin or cis-platin on human myeloid, erythroid, or megakaryocyte lineages (Susick et al., 1993). Threo-dihydrosphingosine potentiated the in vivo antitumor efficacy of cis-platin and adriamycin in 16C mammary tumors and SCCVII carcinomas in mice (Siemann et al., 1993). PKC is activated rapidly and transiently following ionizing radiation. In two human squamous cell carcinoma cell lines exposed to graded doses of X-rays the presence of staurosporine or sangivamycin enhanced cell killing, while H-7 did not (Hallahan et al., 1992). In human ovarian cancer cell lines with intrinsic resistance to cis-platin, PKC activities in the cytosol and membrane were approximately 4- to 5-fold higher than that of sensitive cells. Proliferation of sensitive and resistant cells was inhibited in a dose-dependent manner by TPA. The membrane PKC activities in the cis-platin-sensitive cells were rapidly activated and down-regulated 24 hours after exposure to TPA, while those in the resistant cells were not down-regulated even after exposure to TPA for 24 hours, suggesting that the membrane form of PKC may be involved in the intrinsic resistance. In these cell lines cis-platin sensitivity was reduced by TPA when cellular PKC rose, the sensitivity was increased when cellular PKC decreased (Hirata et al, 1993). In human osteosarcoma U2-OS cells and a cis-platin resistant U2-OS/Pt variant, a 24 hours exposure to TPA caused a potentiation of cis-platin toxicity in sensitive and resistant cells. A short term exposure to TPA did not affect cis-platin cytotoxicity in both cell lines (Perego et al., 1993). Activation of PKC by TPA increased the resistance of PAM 212 keratinocytes transformed by the v-H-ras oncogene to vincristine and adriamycin, while it significantly decreased the resistance in E1a transformed PAM 212 cells to cis-platin. Staurosporine increased the cytotoxicity of vincristine, doxorubicin and cis-platin in the E1a transformed keratinocytes (Sanchez-Prieto et al., 1995). Short term treatment of Chinese hamster ovary cells with TPA increased the resistance to methotrexate in a dose dependent manner. H-7, staurosporine and calphostin C decreased the TPA-induced resistance (Noe and Ciudad, 1995). In C6 glioma cells staurosporine and H-7 increased the sensitivity to irradiation, whereas the PKA inhibitor HA1004 failed to affect the radiosensitivity of these cells (Zhang et al., 1993).

Combination studies with PKCα antisense and standard chemotherapeutic agents (cis-platin, mitomycin-C, vinblastine, estracyt and adriamycin) in nude mice that had been transplanted with a variety of human tumors (breast, prostate, large cell lung and small cell lung carcinomas, and melanomas) were found to be additive or superadditive (Geiger et al, 1998).

Exposure of U251 and LN-Z308 glioma to nitroprusside (a NO-generating agent) resulted in significant cytotoxicity. U343 cells were resistant to the compound and exhibited higher basal levels of PKCα and bcl-2 than U251 and LN-Z308 cells. Introduction of PKCα antisense oligonucleotides into U343 cells decreased the levels of bcl-2 and and also decreased the resistance to nitroprusside. PKCα transfected U251 clones displayed increased PKC acitivty, bcl-2 expression and resistance to nitroprusside (Blackburn et al. 1998).

As described above (section 2.3.3), UCN-01 inhibited the proliferation of human leukemia, breast cancer, malignant glioma, and small cell lung cancer cell lines (Akinaga et al., 1991; Seynaeve et al., 1993; Courage et al., 1995; Pollack et al., 1996). The compound exhibited antitumor activity against human tumor xenografts in nude mice (Akinaga et al., 1991). Clinical trials with the compound are ongoing (Lush et al., 1997). UCN-01 enhanced the antitumor activity of mitomycin C in human A431 epidermoid carcinoma, human xenografted colon carcinoma Co-3 and murine sarcoma 180 cells in vitro and in vivo (Akinaga et al, 1993). UCN-01 also blocked the proliferation of glioma cells in vitro and in vivo and potentiated the effects of BCNU and cis-platin (Pollak et al., 1996). In an investigation by Husain et al., (1997) in all cell lines studied UCN-01 was effective as a cytotoxic agent alone and in combination with cis-platin. The combination of UCN-01 plus cis-platin was effective in increasing the cytostatic and cytotoxic effects of irradiation at 4 Gy (Sommers and Alfieri, 1998). In Chinese hamster ovary cells UCN-01 abrogated the G2 arrest induced by the DNA-damaging agent cis-platin. UCN-01 concentrations that resulted in abrogation of the cis-platin-induced G2 arrest also enhanced cis-platin-induced cytotoxicity. UCN-01 enhanced cis-platin cytotoxicity up to 60-fold and reduced by 3-fold the concentration of cis-platin required to kill 90% of the cells. The concentrations of UCN-01 required for this enhancement have been shown to be well tolerated in animal models, suggesting that this combination may represent an effective strategy for enhancing cis-platin-based chemotherapeutic regimens (Bunch and Eastman, 1996).

Pretreatment of human T24 bladder or human colorectal adenocarcinoma cells with 5-FU followed by CGP 41251 showed a synergistic drug interaction in both cell lines (Fabbro et al., 1999). CGP 41251 or 5-FU were ineffective as a single agent against COLO 205 tumors up to 200 mg/kg/day p.o. and 75 mg/kg/week i. v., respectively. Combination of both compounds at these concentrations showed significant antitumor activity in this tumor model (Fabbro et al., 1999).

5.2 PKC Activation Enhances Sensitivity to Antitumor Drugs

Short term treatment with TPA sensitized human 2008 ovarian carcinoma cells to cis-platin. This sensitization disappeared completely by seven hours after treatment, indicating that not inhibition, but activation of PKC sensitizes 2008 cells to the antiproliferative activity of cis-platin (Isonishi et al., 1990). Pretreatment of HeLa cells with TPA or PdBu caused a 9-fold increase in cellular sensitivity to cis-platin and 2.5-fold to melphalan, but had now effect on the antiproliferative activity of bleomycin, adriamycin, vincristine, or mitomycin C. The sensitization of HeLa cells by TPA was associated with a 6-fold stimulation of PKC activation and a concentration- and time-dependent increase in cellular platinum content. (Basu et al. 1990). PKC activity was found to be decreased significantly in cis-platin-resistant human small cell lung H69/CP cancer cells compared to the drug-sensitive variant. A similar reduction in PKC activity was noted in ovarian carcinoma 2008 cells that were resistant to cis-platin. A modest decrease in PKC activity was also observed in etoposide-resistant H69 cells but not in taxol-resistant H69 cells or bleomycin-resistant human head and neck carcinoma A-253 cells (Basu et al., 1996), indicating that reduced PKC activity leads to decreased sensitivity in this system.

5.3 PKC Isoenzymes Involved in the Modulation of Antiproliferative Activity

TPA induced pleiotropic resistance against antitumor drugs in human colon cancer cells. The resistance was triggered by the activation of PKCα. PKCα is expressed abundantly in surgical specimens of human colon cancer, indicating that PKCα-mediated drug resistance may contribute to the intrinsic resistance of clinical colon cancer (O´Brian et al., 1995). Downmodulation of PKCα, but not of PKCβ, with antisense oligonucleotides sensitized human colon carcinoma cells to mitomycin C, 5-FU and vincristine (Chakrabarty and Huang, 1996). H69 human small cell lung cancer cells expressed PKCα, β, δ, ζ, ι and μ. A decrease in PKCα and β and an increase in PKCδ were observed in a cis-platin-resistant subline. The abundance of PKCζ or ι was unaffected. H69 cells resistant to etoposide also displayed a reduction in PKCβ and an increase in PKCδ. Taxol-resistant H69 cells showed no alteration in the expression of any of the PKC isozymes (Basu et al., 1996). R6 cells stably transfected with PKCε were shown to prevent cis-platin-induced apoptosis and protected cells against cis-platin cytotoxicity (Basu and Cline, 1995). PKCε was overexpressed in human lung A549 adenocarcinoma cells following irradiation (Kim et al., 1992). The parental ovarian carcinoma 2008 cells expressed the PKCα, ε, and ζ isoforms. In resistant 2008 cells PKCα decreased significantly, whereas the amount of PKCε increased moderately, with no alteration in the PKCζ content. Therefore, the authors concluded that a reduction in PKCα and/or an increase in PKCε expression may be associated with the drug-resistant phenotype (Basu and Weixel, 1995). Antisense oligodeoxynucleotides against PKCε down-regulated the PKCε level, blocked drug-induced translocation, and reduced cis-platin-mediated cytotoxicity 3-fold compared to that of sense-treated cells. Antisense PKCε also decreased SKBR-3 cell sensitivity to carboplatin but not to adriamycin and taxol (Ohmori and Arteaga, 1998). Therefore, the effects of PKCε may be cell type specific and also drug specific.

5.4 Remarks and Conclusions

Contradicting results on the effects of PKC on drug resistance in different publications may have several reasons. One reason may be the use clonogenic assays or short term assays. For example, treatment of mouse embryo fibroblasts from wild-type ($p53^{+}/^{+}$) and from p53 knockout mice ($p53^{-}/^{-}$) with etoposide led to increased sensitivity of wild-type expressing cells tested in a short-term XTT-assay. However, no difference in sensitivity was observed following etoposide treatment if tested in a clonogenic assay (Brown and Wouters, 1999). In a comparison between matched $p53^{+}/^{+}$ and $p53^{-}/^{-}$ mouse cells, five investigations found $p53^{-}/^{-}$ cells more radioresistant, nine found no difference, and three found $p53^{-}/^{-}$ more sensitive than $p53^{+}/^{+}$ cells (Brown and Wouters, 1999).

UV-2237MM cells exhibited high sensitivity to adriamycin in vitro. If these cells were transplanted into different organs of mice, tumors growing in the subcutis and the spleen were sensitive, tumors growing in the lung were resistant to adriamycin. If tumor cells from the lung were isolated and grown in vitro they were sensitive again. Nearly identical levels of PKC activity in tumors growing in the subcutis, spleen and lung were observed, indicating that PKC activity levels did not account for the different responses to adriamycin (Staroselsky et al., 1990).

The influence of PKC on cellular sensitivity or resistance to cis-platin, adriamycin or irradiation may arise from the status of phosphorylation of raf-1, bcl-2, IκB; IκB kinase β, and many others (Grant and Jarvis, 1996; Schönwasser et al., 1998; Lallena et al., 1999). It has been shown that PKCα phosphorylates raf-1 (Kolch et al, 1993) and bcl-2, leading to resistance (Ruvolo et al, 1998). Raf-1 also interacts with the apoptosis-preventing bcl-2 (Blagosklonny et al, 1997). Phorbol ester treatment of quiescent Swiss 3T3 cells led to cell proliferation, a response thought to be mediated by PKC. In addition to activation of raf-1 by PKCα, it was described that this isoenzyme induced also raf desensitization to prevent further raf stimulation by growth factors (Schönwasser et al., 1998). In 15 human cell lines high cyclin D1 expression was related to cis-platin resistance but had no relationship with radiation responsiveness, whereas high c-raf-1 expression, although related to radiosensitivity has no relationship with cis-platin responsiveness (Warenius et al., 1996). When transferred from monolayer to three-dimensional culture, a consistent upregulation (up to 15-fold) of the cyclin-dependent kinase inhibitor p27^{KIP1} protein was observed in a panel of mouse and human carcinoma cell lines which was accompagnied by resistance to antitumor drugs and radiation, implicating that p27^{KIP1} is a regulator of drug resistance in solid tumors (StCroix et al., 1996; StCroix and Kerbel, 1997). These are indications that, in addition to PKC, many factors influence drug resistance and cell death.

Treatment of cultured mammalian cells with serum growth factors and activators of PKC, led to induction of metallothionein mRNA. One of the required steps in the signal transduction pathways triggered by these agents ending in metallothionein gene induction, appears to be the activation of protein kinase C (Imbra and Karin, 1987). Resistance to cis-platin may be due to induction of metallothionein genes by PKC and inhibition of PKC may sensititze cells by inhibition of metallothionein gene expression (Yu et al., 1997).

6 Summary

PKC isoenzymes were found to be involved in proliferation, antitumor drug resistance and apoptosis. Therefore, it has been tried to exploit PKC as a target for antitumor treatment. PKCα activity was found to be elevated, for example, in breast cancers and malignant gliomas, whereas it seems to be underexpressed in many colon cancers. So it can be expected that inhibition of PKC activity will not show similar antitumor activity in all tumors. In some tumors it seems to be essential to inhibit PKC to reduce growth. However, for inhibition of tumor proliferation it may be an advantage to induce apoptosis. In this case an activation of PKCδ should be achieved. The situation is complicated by the facts that bryostatin leads to the activation of PKC and later to a downmodulation and that the PKC inhibitors available to date are not specific for one PKC isoenzyme. For these reasons, PKC modulation led to many contradicting results. Despite these problems, PKC modulators such as miltefosine, bryostatin, safingol, CGP41251 and UCN-01 are used in the clinic or are in clinical evaluation. The question is whether PKC is the major or the only target of these compounds, because they also interfere with other targets.

PKC may also be involved in apoptosis. Oncogenes and growth factors can induce cell proliferation and cell survival, however, they can also induce apoptosis, depending on the cell type or conditions in which the cells or grown. PKC participates in these signalling pathways and cross-talks. Induction of apoptosis is also dependent on many additional factors, such as p53, bcl-2, mdm2, etc. Therefore, there are also many contradicting results on PKC modulation of apoptosis. Similar controversial data have been reported about MDR1-mediated multidrug resistance. At present it seems that PKC inhibition alone without direct interaction with PGP will not lead to successful reversal of PGP-mediated drug efflux. One possibility to improve chemotherapy would be to combine established antitumor drugs with modulators of PKC. However, here also very contrasting results were obtained. Many indicate that inhibition, others, that activation of PKC enhances the antiproliferative activity of anticancer drugs. The problem is that the exact functions of the different PKC isoenzymes are not clear at present. So further investigations into the role of PKC isoenzymes in the complex and interacting signalling pathways are essential. It is a major challenge in the future to reveal whether modulation of PKC can be used for the improvement of cancer therapy.

References

Accornero P, Radrizzani M, Care A, Mattia G, Chiodoni C, Kurrle R, Colombo MP (1998) HIV/gp12 and PMA/ionomycin induced apoptosis but not activation induced cell death require PKC for Fas-L upregulation. FEBS Lett 436:467-465

Acs P, Wang QJ, Bogi K, Marquez AM, Lorenzo PS, Biro T, Szallasi Z, Mushinski JF, Blumberg PM (1997) Both the catalytic and regulatory domains of protein kinase C chimeras modulate the proliferative properties of NIH 3T3 cells. J Biol Chem 272:28793-28799

Adams LM, Dykes D, Harrison SD, Saleh J, Saah L (1993) Combined effect of the chemopotentiator SPC-100270, a protein kinase C (PKC) inhibitor, and doxorubicin (dox) or cisplatin (cis) on murine isografts and human tumor xenografts. Proc Am Assoc Cancer Res 34:410, 2448

Aebi S, Kurdi-Haidar B, Gordon R, Cenni B, Zheng H, Fink D, Christen RD, Boland RC, Koi N, Fishel R, Howell SB (1996) Loss of DNA mismatch repair in acquired resistance to cisplatin. Cancer Res 56:3087-3090

Aftab DT, Yang JM, HaitWN (1994) Functional role of phosphorylation of the multidrug transporter (P-glycoprotein) by protein kinase C in multidrug-resistant MCF-7 cells. Oncol Res 6:59-70

Ahmad S, Trepel JB, Ohno S, Suzuki K, Tsuruo T, Glazer RI (1992) Role of protein kinase C in the modulation of multidrug resistance: expression of the atypical gamma isoform of protein kinase C does not confer increased resistance to doxorubicin. Mol Pharmacol 42:1004-1009

Ahmad S, Glazer RI (1993) Expression of the antisense cDNA for protein kinase C alpha attenuates resistance in doxorubicin-resistant MCF-7 breast carcinoma cells. Mol Pharmacol 43: 858-862

Ahmad S, Mineta T, Martuza RL, Glazer RI (1994) Antisense expression of protein kinase C alpha inhibits the growth and tumorigenictity of human glioblastoma cells. Neurosurgery 35:904-908

Ahmed S, Lee J, Kozma R, Best A, Monfries C, Lim LA (1993) A novel functional target for tumor-promoting phorbol esters and lysophosphatidic acid. The p21rac-GTPase activating protein n-chimaerin. J Biol Chem 268:10709-10712

Ahmed MM, Venkatasubbarao K, Fruitwala SM, Muthukkumar S, Wood DP Jr, Sells SF, Mohiuddin M, Rangnekar VM (1996) EGR-1 induction is required for maximal radiosensitivity in A375-C6 melanoma cells. J Biol Chem 271:29231-29237

Ahn CH, Kong JY, Choi WC, Wang MS (1996) Selective inhibition of the effects of phorbol ester on doxorubicin resistance and P-glycoprotein by the protein kinase C inhibitor 1-(5-isoquinolinesulfonyl)-2-methylpiperazine (H7) in multidrug-resistant MCF-7/Dox human breast carcinoma cells. Biochem Pharmacol 52:393-399

Akinaga S, Gomi K, Morimoto M, Tamaoki T, Okabe M (1991) Antitumor activity of UCN-01, a selective inhibitor of protein kinase C, in murine and human tumor models. Cancer Res 51:4888-4892

Akinaga S, Nomura K, Gomi K, Okabe M (1993) Enhancement of antitumor activity of mitomycin C in vitro and in vivo by UCN-01, a selective inhibitor of protein kinase C. Cancer Chemother Pharmacol 32:183-189

Akiyama T, Yoshida T, Tsujita T, Shimizu M, Mizukami T, Okabe M, Akinaga S (1997) G1 phase accumulation induced by UCN-01 is associated with dephosphorylation of Rb and CDK2 proteins as well as induction of CDK inhibitor p21/Cip1/WAF1/Sdi1 in p53-mutated human epidermoid carcinoma A431 cells. Cancer Res 57:1495-1501

Alvaro V, Touraine P, Raisman Vozari R, Bai-Grenier F, Birman P, Joubert D (1992) Protein kinase C activity and expression in normal and adenomatous human pituitaries. Int J Cancer 50:724-730

Alvaro V, Levy L, Dubray C, Roche A, Peillon F, Querat B, Joubert D (1993) Invasive human pituitary tumors express a point-mutated alpha-protein kinase C. J Clin Endocrinol Metab 77:1125-1129

Alvaro V, Prevostel C, Joubert D, Slosberg E, Weinstein IB (1997) Ectopic expression of a mutant form of PKCalpha originally found in human tumors: aberrant subcellular translocation and effects on growth control. Oncogene 14:677-685

Aquino A, Hartman KD, Knode MC, Grant S, Huang KP, Niu CH, Glazer RI (1988) Role of protein kinase C in phosphorylation of vinculin in adriamycin resistant HL 60 leukemia cells. Cancer Res 48:3324-3329

Asada A, Zhao Y, Kondo S, Iwata M (1998) Induction of thymocyte apoptosis by Ca^{2+}-independent protein kinase C (nPKC) activation and its regulation by calcineurin activation. J Biol Chem 273:28392-28398

Ask A, Persson L, Rehnholm A, Frostesjo L, Holm I, Heby O (1993) Development of resistance to hydroxyurea during treatment of human myelogenous leukemia K562 cells with alpha-difluoromethylornithine as a result of coamplification of genes for ornithine decarboxylase and ribonucleotide reductase R2 subunit. Cancer Res 53:5262-5268

Assert R, Kruis W, Hardt M, Fischbach W, Schatz H, Pfeiffer A (1993) Expression of protein-kinase C (PKC)-a and b2 mRNA in normal mucosa, andomas and cell

lines and growth arrest by PKC in differentiated colon T84 cells. Gastroenterology 89 (Suppl. 1), A2521

Assert R, Kötter R, Bisping G, Scheppach W, Stahlnecker E, Müller KM, Dusel G,
Schatz H, Pfeiffer A (1999) Anti-proliferative activity of protein kinase C in apical
compartments of human colonic crypts: evidence for a less activated protein
kinase C in small adenomas. Int J Cancer 80:47-53

Azuma Y, Onishi Y, Sato Y, Kizaki H (1993) Induction of mouse thymocyte apoptosis by inhibitors of tyrosine kinases is associated with dephosphorylation of nuclear proteins. Cell Immunol 152:271-278

Azzi A, Boscoboinik D, Hensey C (1992) The protein kinase C and protein kinase C
related gene families. Eur J Biochem 208:547-557

Baltuch GH, Dooley NP, Couldwell WT, Yong VW (1993) Staurosporine differentially inhibits glioma versus non-glioma cell lines. J Neurooncol16:141-147

Baltuch GH, Dooley NP, Rostworowski KM, Villemure JG, Yong VW (1995) Protein
kinase C isoform alpha overexpression in C6 glioma cells and its role in cell proliferation. J Neurooncol 24:241-250

Bamberger AM, Bamberger CM, Wald M, Kratzmeier M, Schulte HM (1996) Protein
kinase C (PKC) isoenzyme expression pattern as an indicator of proliferative activity in uterine tumor cells. Mol Cell Endocrinol 123:81-88

Basu A, Teicher BA, Lazo JS (1990) Involvement of protein kinase C in phorbol
ester-induced sensitization of HeLa cells to cis-diamminedichloroplatinum(II) J
Biol Chem 265:8451-8457

Basu A (1993) The potential of protein kinase C as a target for anticancer treatment.
Pharmac. Ther. 59:257-280

Basu A, Cline JS (1995) Oncogenic transformation alters cisplatin induced apoptosis
in rat embryo fibroblasts. Int J Cancer 63:597-603

Basu A, Weixel KM (1995) Comparison of protein kinase C activity and isoform
expression in cisplatin-sensitive and –resistant ovarian carcinoma cells. Int J
Cancer 62:457-460

Basu A, Weixel K, Saijo N (1996) Characterization of the protein kinase C signal
transduction pathway in cisplatin-sensitive and –resistant human small cell lung
carcinoma cells. Cell Growth Differ 7:1507-1512

Bates SE, Lee JS, Dickstein B, Spolyar M, Fojo AT (1993) Differential modulation of
P-glycoprotein transport by protein kinase inhibition. Biochemistry. 32 9156-64

Batlle E, Verdu J, Dominguez D, del Mont Llosas M, Diaz V, Loukili N, Paciucci R,
Alameda F, de Herreros AG (1998) Protein kinase C-alpha activity inversely
modulates invasion and growth of intestinal cells. J Biol Chem 273:15091-15098

Baxter G, Oto E, Daniel-Issakani S, Strulovici B (1992) Constitutive presens of a
catalytic fragment of protein kinase Ce in a small cell lung carcinoma cell line. J
Biol Chem 267:1910-1917

Beck J, Handgretinger R, Klingebiel T, Dopfer R, Schaich M, Ehninger G, Nietham-
mer D, Gekeler V (1996) Expression of PKC isozyme and MDR-associated genes
in primary and relapsed state AML. Leukemia. 10:426-433

Beck J, Bohnet B, Brugger D, Bader P, Dietl J, Scheper RJ, Kandolf R, Liu C, Nieth-
ammer D, Gekeler V (1998) Multiple gene expression analysis reveals distinct
differences between G2 and G3 stage breast cancers, and correlations of PKC eta
with MDR1, MRP and LRP gene expression. Br J Cancer 77:87-91

Begemann M, Kashimawo SA, Choi YJA, Kim S, Christiansen KM, Duigou G, Mueller
M, Schieren I, Gosh S, Fabbro D, Lampen NM (1996) Inhibition of glioblastomas

by CGP 41251, an inhibitor of protein kinase C, and by a phorbol ester tumor promoter. Clin Cancer Res 2:1018-1030

Bell SE, Quinn DM, Kellett GL, Warr JR (1998) 2-Deoxy-D-glucose preferentially kills multidrug-resistant human KB carcinoma cell lines by apoptosis. Br J Cancer 78:1464-1470

Beltran PJ, Fan D, Fidler IJ, O'Brian CA (1997) Chemosensitization of cancer cells by the staurosporine derivative CGP 41251 in association with decreased P-glycoprotein phosphorylation. Biochem Pharmacol 53:245-247

Benzil DL, Finkelstein SD, Epstein MH, Finch PW (1992) Expression pattern of alpha-protein kinase C in human astrocytomas indicates a role in malignant progression. Cancer Res 52:2951-2956

Berdel WE (1991) Membrane-interactive lipids as experimental anticancer drugs. Br J Cancer 64:208-211

Bergman PJ, Gravitt KR, Ward NE, Beltran P, Gupta KP, O´Brian CA (1997) Potent inuduction of human colon cancer cell uptake of chemotherapeutic drugs by N-myristoylated protein kinase C-alpha (PKC-alpha) pseudosubstrate peptides through a P-glycoprotein-independent mechanisms. Invest New Drugs 15:311-318

Berkovic D, Berkovic K, Fleer EA, Eibl H, Unger C (1994) Inhibition of calcium-dependent protein kinase C by hexadecylphosphocholine and 1-O-octadecyl-2-O-methyl-rac-glycero-3-phosphocholine do not correlate with inhibition of proliferation of HL60 and K562 cell lines. Eur J Cancer 30A:509-515

Berkow RL, Kraft AS (1989) Bryostatin, a nonphorbol macrocyclic lactone activates intact human polymorphonuclear leukocytes and binds to the phorbol ester receptor. Biochem Biophys Res Commun 131:1109-1115

Bernards R (1991) N-myc disrupts protein kinase C-mediated signal transduction in neuroblastoma. EMBO J 10:1119-1125

Berra E, Diaz-Meco MT, Dominguez I, Municio MM, Sanz L, Lozano J, Chapkin RS, Moscat J (1993) Protein kinase C zeta isoform is critical for mitogenic signal transduction. Cell 74:555-563

Berra E, Municio MM, Sanz L, Frutos S, Diaz-Meco MT, Moscat J (1997) Positioning atypical protein kinase C isoforms in the UV induced apoptotic signaling cascade. Mol Cell Biol 17:4346-4354

Bhalla K, Ibrado AM, Tourkina E, Tang C, Grant S, Bullock G, Huang Y, Ponnathpur V, Mahoney ME (1993) High-dose mitoxantrone induces programmed cell death or apoptosis in human myeloid leukemia cells. Blood 82(10):3133-3140

Blackburn RV, Galoforo SS, Berns CM, Motwani NM, Corry PM, Lee YJ (1998) Differential induction of cell death in human glioma cell lines by sodium nitroprusside. Cancer 82:1137-1145

Blagosklonny MV, Giannakakou P, el-Deiry WS, Kingston DG, Higgs PI, Neckers L, Fojo T (1997) Raf-1/bcl-2 phosphorylation: a step from microtubule damage to cell death. Cancer Res 57: 130-135

Blagosklonny MV (1999) A node between proliferation, apoptosis, and growth arrest. BioEssays 21:704-709

Blake RA, Garcia-Paramio P, Parker PJ, Courtneidge SA (1999) Src promotes PKCδ degradation. Cell Growth Diff 10:231-241

Blattner C, Tobiasch E, Litfen M, Rahmsdorf HJ, Herrlich P (1999) DNA damage induced p53 stabilization: no indication for an involvement of p53 phosphorylation. Oncogene 18:1723-1732

Blobe GC, Sachs CW, Khan WA, Fabbro D, Stabel S, Wetsel WC, Obeid LM, Fine RL, Hannun YA (1993) Selective regulation of expression of protein kinase C (PKC) isoenzymes in multidrug-resistant MCF-7 cells. Functional significance of enhanced expression of PKCα. J Biol Chem 268:658-664

Blobe GC, Obeid LM, Hannun YA (1994) Regulation of protein kinase C and role in cancer biology. Cancer Metastasis Rev 13:411-431

Blumberg PM (1991) Complexities of the protein kinase C pathway Mol Carcinog 4:339-344

Borgatti P, Mazzoni M, Carini C, Neri LM, Marchisio M, Bertolaso L, Previati M, Zauli G, Capitani S (1996) Changes of nuclear protein kinase C activity and isotype composition in PC12 cell proliferation and differentiation. Exp Cell Res 224:72-78

Borner C, Wyss R, Regazzi R, Eppenberger U, Fabbro D (1987) Immunological quantitation of phospholipid/Ca^{2+}-dependent protein kinase of human mammary carcinoma cells: inverse relationship to estrogen receptors. Int J Cancer 40:344-348

Borner C, Filipuzzi I, Weinstein IB, Imber R (1991) Failure of wild-type or a mutant form of protein kinase C-alpha to transform fibroblasts. Nature 353:78-80

Brachwitz H, Vollgraf C (1995) Analogs of alkyllysophospholipids: chemistry, effects on the molecular level and their consequences for normal and malignant cells. Pharmacol Ther 66:39-82

Brooks G, Wilson RE, Dooley TP, Goss MW, Hart IR (1991) Protein kinase C down-regulation, and not transient activation, correlates with melanocyte growth. Cancer Res 51:3281-3288

Brown MJ, Wouters BG (1999) Apoptosis, p53, and tumor cell sensitivity to anticancer agents. Cancer Res 59:1391-1399

Buchner K (1995) Protein kinase C in the transduction of signals toward and within the cell nucleus. Eur J Biochem 228:211-221

Budworth J, Davis R, Malkhandi J, Gant TW, Ferry DR, Gescher A (1996) Comparison of staurosporine and four analogues: their effects on growth, rhodamine 123 retention and binding to P-glycoprotein in multidrug-resistant MCF-7/Adr cells. Br J Cancer 73:1063-1068

Budworth J, Gant TW, Gescher A (1997) Co-ordinate loss of protein kinase C and multidrug resistance gene expression in revertant MCF-7/Adr breast carcinoma cells. Br J Cancer 75:1330-1335

Bunch RT, Eastman A (1996) Enhancement of cisplatin-induced cytotoxicity by 7-hydroxystaurosporine (UCN-01), a new G2-checkpoint inhibitor. Clin Cancer Res 2:791-797

Burger C, Wick M, Muller R (1994) Lineage-specific regulation of cell cycle gene expression in differentiating myeloid cells. J Cell Sci 107:2047-2054

Bush RS, Jenkin RD, Allt WE, Beale FA, Bean H, Dembo AJ, Pringle JF (1978) Definitive evidence for hypoxic cells influencing cure in cancer therapy. Br J Cancer 37:(Suppl III), 302-306

Cacace AM, Guadagno SN, Krauss RS, Fabbro D, Weinstein IB (1993) The epsilon isoform of protein kinase C is an oncogene when overexpressed in rat fibroblasts. Oncogene 8:2095-2104

Cai H, Smola U, Wixler V, Eisenmann-Tappe I, Diaz-Meco MT, Moscat J, Rapp U, Cooper GM (1997) Role of diacylglycerol-regulated protein kinase C isotypes in growth factor activation of the Raf-1 protein kinase. Mol Cell Biol 17:732-741

Caloca MJ, Fernandez N, Lewin NE, Ching D, Modali R, Blumberg PM, Kazanietz MG (1997) Beta2-chimaerin is a high affinity receptor for the phorbol ester tumor promoters. J Biol Chem 272:26488-26496

Capiati DA, Limbozzi F, Tellez-Inon MT, Boland RL (1999) Evidence on the participation of protein kinase C alpha in the proliferation of cultured fibroblasts. J Cell Biochem 74:292-300

Caravatti G, Meyer T, Fredenhagen A, Trinks U, Mett H, Fabbro D (1994) Inhibitory activity and selectivity of staurosporine derivatives towards protein kinase C. Bioorg Med Chem Lett 4:399-404

Carey I, Noti JD (1999) Isolation of protein kinase C-alpha-regulated cDNAs associated with breast tumor aggressiveness by differential mRNA display. Int J Oncol 14:951-956

Carnero A, Liyanage M, Stabel S, Lacal JC (1995) Evidence for different signalling pathways of PKC zeta and ras-p21 in Xenopus oocytes. Oncogene 11:1541-1547

Cartee L, Kucera GL (1998) Gemcitabine induces programmed cell death and activates protein kinase C in BG-1 human ovarian cancer cells. Cancer Chemother Pharmacol 41:403-412

Cartee L, Kucera GL, Nixon JB (1998) The effects of gemcitabine and TPA on PKC signaling in BG-1 human ovarian cancer cells. Oncol Res 10:371-377

Castagna M, Takai Y, Kaibuchi K, Sano K, Kikkawa U, Nishizuka Y (1982) Direct activation of calcium-activated, phospholipid-dependent protein kinase by tumor-promoting phorbol esters. J Biol Chem 257:7847-7851

Castro AF, Horton JK, Vanoye CG, Altenberg GA (1999) Mechanism of inhibition of P-glycoprotein-mediated drug transport by protein kinase C blockers. Biochem Pharmacol 58:1723-1733

Chakrabarty S, Huang S (1996) Modulation of chemosensitivity in human colon carcinoma cells by downregulating protein kinase C alpha expression. J Exp Ther Oncol 1:218-221

Chambers TC, McAvoy EM, Jacobs JW, Eilon G (1990a) Protein kinase C phosphorylates P-glycoprotein in multidrug-resistant human KB carcinoma cells. J Biol Chem 265:7679-7686

Chambers TC, Chalikonda I, Eilon G (1990b) Correlation of protein kinase C translocation, P-glycoprotein phosphorylation and reduced drug accumulation in multidrug resistant human KB cells. Biochem Biophys Res Commun 169:253-259

Chambers TC, Zheng B, Kuo JF (1992) Regulation by phorbol ester and protein kinase C inhibitors, and by a protein phosphatase inhibitor (okadaic acid) of P-glycoprotein phosphorylation and relationship to drug accumulation in multidrug-resistant human KB cells. Mol Pharmacol 41:1008-1015

Chambers TC, Pohl J, Raynor RL, Kuo JF (1993) Identification of specific sites in human P-glycoprotein phosphorylated by protein kinase C. J Biol Chem, 268:4592-4595

Chambers TC, Pohl J, Glass DB, Kuo JF (1994) Phosphorylation by protein kinase C and cyclic AMP-dependent protein kinase of synthetic peptides derived from the linker region of human P-glycoprotein. Biochem J 299: 309-315

Chaudhary PM, Roninson IB (1992) Activation of MDR1 (P-glycoprotein) gene expression in human cells by protein kinase C agonists. Oncology Research 4:281-290

Chaudhary PM, Roninson IB (1993) Induction of multidrug resistance in human cells by transient exposure to different chemotherapeutic drugs. J Natl Cancer Inst 85:532-539

Chauhan A, Chauhan VPS, Brockerhoff H, Wisniewski HM (1991) Action of amyloid beta-protein on protein kinase C activity. Life Sci 49:1555-1562

Chen PL, Chen CF, Chen Y, Xiao J, Sharp ZD, Lee WH (1998) The BRC repeats in BRCA2 are critical for RAD51 binding and resistance to methyl methanesulfonate treatment. Proc Natl Acad Sci USA 95:5287-5292

Chen CY, Faller DV (1999) Selective inhibition of protein kinase C isoenzymes by Fas ligation. J Biol Chem 274:15320-15328

Chen JJ, Silver D, Cantor S, Livingston DM, Scully R (1999) BRCA1, BRCA2, and Rad51 operate in a common DNA damage response pathway. Cancer Res 59(7 Suppl):1752s-1756s

Cheng AL, Chuang SE, Fine RL, Yeh KH, Liao CM, Lay JD, Chen DS (1998) Inhibition of the membrane translocation and activation of protein kinase C, and potentiation of doxorubicin-induced apoptosis of hepatocellular carcinoma cells by tamoxifen. Biochem Pharmacol 55:523-531

Chida K, Kato N, Kuroki T (1986) Down regulation of phorbol diester receptors by proteolytic degradation of protein kinase C in a cultured cell line of fetal rat skin keratinocytes. J Biol Chem 261:13013-13018

Chmura SJ, Nodzenski E, Weichselbaum RR, Quintans J (1996a) Protein kinase C inhibition induces apoptosis and ceramide production through activation of a neutral sphingomyelinase. Cancer Res 56:2711-2714

Chmura SJ, Nodzenski E, Crane MA, Virudachalam S, Hallahan DE, Weichselbaum RR, Quintans J (1996b) Cross-talk between ceramide and PKC activity in the control of apoptosis in WEHI-231. Adv Exp Med Biol 406:39-55

Cho Y, Klein MG, Talmage DA (1998) Distinct functions of protein kinase Calpha and protein kinase Cbeta during retinoic acid-induced differentiation of F9 cells. Cell Growth Differ 9:147-154

Choi PM, Tchou-Wong KM, Weinstein IB (1990) Overexpression of protein kinase C in HT29 colon cancer cells causes growth inhibition and tumor suppression. Mol Cell Biol 10:4650-4657

Chu G (1994) Cellular responses to cisplatin. The roles of DNA-binding proteins and DNA repair. J Biol Chem 269:787-790

Ciocca DR, Oesterreich S, Chamness GC, McGuire WL, Fuqua SAW (1993) Biological and clinical implications of heat shock protein 27000 (Hsp27: a review. J Natl Cancer Inst 85:1558-1570

Civoli F, Daniel LW (1998) Quaternary ammonium analogs of ether lipids inhibit the activation of protein kinase C and the growth of human leukemia cell lines. Cancer Chemother Pharmacol 42:319-326

Clavel M, Mauriac L, Vennin P, Namer M, Maugard-Louboutin C, Goupil A, Veyret C, Sindermann H, David M (1992) Topically applied miltefosine (hexadecylphosphocholine) in cutaneous relapses and/or skin metastasized breast cancer: a French multicentric study. Ann Oncol 3 (Suppl. 1): 65

Cloud-Heflin BA, McMasters RA, Osborn MT, Chambers TC (1996) Expression, subcellular distribution and response to phorbol esters of protein kinase C (PKC) isozymes in drug-sensitive and multidrug-resistant KB cells evidence for altered regulation of PKC-alpha. Eur J Biochem 239:796-804

Cole SP, Sparks KE, Fraser K, Loe DW, Grant CE, Wilson GM, Deeley RG (1994) Pharmacological characterization of multidrug resistant MRP-transfected human tumor cells. Cancer Res 54:5902-5910

Coppock DL, Buffolino P, Kopman C, Nathanson L (1995) Inhibition of the melanoma cell cycle and regulation at the G1/S transition by 12-O-tetradecanoylphorbol-13-acetate (TPA) by modulation of CDK2 activity. Exp Cell Res 221:92-102

Cornford, P, Evans J, Dodson A, Parsons K, Woolfenden A, Neoptolemos J, Foster CS (1999) Protein kinase C isoenzyme patterns characteristically modulated in early prostate cancer. Am J. Pathol 154:137-144

Couldwell WT, Antel JP, Apuzzo ML, Yong VW (1990) Inhibition of growth of established human glioma cell lines by modulators of the protein kinase-C system. J Neurosurg 73:594-600

Couldwell WT, Antel JP, Yong VW (1992) Protein kinase C activity correlates with the growth rate of malignant gliomas: Part II. Effects of glioma mitogens and modulators of protein kinase C. Neurosurgery 31:717-724

Couldwell WT, Hinton DR, He S, Chen TC, Sebat I, Weiss MH, Law RE (1994) Protein kinase C inhibitors induce apoptosis in human malignant glioma cell lines. FEBS Lett 345:43-46

Courage C, Budworth J, Gescher A (1995) Comparison of ability of protein kinase C inhibitors to arrest cell growth and to alter cellular protein kinase C localisation. Br J Cancer 71:697-704

Courage C, Snowden R, Gescher A (1996) Differential effects of staurosporine analogs on cell cycle, growth and viability in A549 cells. Br J Cancer 74:1199-1205

Courage C, Bradder SM, Jones T, Schultze-Mosgau MH, Gescher A (1997) Characterisation of novel human lung carcinoma cell lines selected for resistance to anti-neoplastic analogues of staurosporine. Int J Cancer 73:763-768

Craven PA, DeRubertis FR (1988) Role of activation of protein kinase C in the stimulation of colonic epithelial proliferation by unsaturated fatty acids. Gastroenterology 95:676-85

Craven PA, DeRubertis FR (1992) Alterations in protein kinase C in 1,2-dimethylhydrazine induced colonic carcinogenesis. Cancer Res 52:2216-2221

Crespo P, Mischak H, Gutkind JS (1995) Overexpression of mammalian protein kinase C-zeta does not affect the growth characteristics of NIH 3T3 cells. Biochem Biophys Res Commun 213: 266-272

Cui Y, Konig J, Buchholz JK, Spring H, Leier I, Keppler D (1999) Drug resistance and ATP-dependent conjugate transport mediated by the apical multidrug resistance protein, MRP2, permanently expressed in human and canine cells. Mol Pharmacol 55:929-937

Czendlik C, Graf P (1996) Safety, tolerability and pharmacokinetics of CGP 41251, a protein kinase C inhibitor (phase I study). Ann Oncol 7, Suppl 1: 264.

Dale IL, Gescher A (1989) Effects of activators of protein kinase C, including bryostatins 1 and 2, on the growth of A549 human lung carcinoma cells. Int J Cancer 43:158-163

Dale IL, Bradshaw TD, Gescher A, Pettit GR (1989) Comparison of effects of bryostatins 1 and 2 and 12-O-tetradecanoylphorbol-13-acetate on protein kinase C activity in A549 human lung carcinoma cells. Cancer Res 49:3242-3245

Darbon JM, Valette A, Bayard F (1986) Phorbol esters inhibit the proliferation of MCF-7 cells. Possible implication of protein kinase C. Biochem Pharmacol 35:2683-2686

Das KC, Guo X, White CW (1998) Protein kinase Cδ-dependent induction of Manganese superoxide dismutase gene expression by microtubule-active anticancer drugs. J Biol Chem 273:34639-34645

Datta R, Kojima H, Yoshida K, Kufe D (1997) Caspase 3 mediated cleavage of protein kinase C theta in induction of apoptosis. J Biol Chem 272:20317-20320

Davies R, Budworth J, Riley J, Snowden R, Gescher A, Gant TW (1996) Regulation of P-glycoprotein 1 and 2 gene expression and protein activity in two MCF-7/Dox cell line subclones. Br J Cancer 73: 307-315

Deacon EM, Pongracz J, Griffiths G, Lord JM (1997) Isoenzymes of protein kinase C: differential involvement in apoptosis and pathogenesis. J Clin Pathol: Mol Pathol 50:124-131

Dean N, McKay R, Miraglia L, Howard R, Cooper S, Giddings J, Nicklin P, Meister L, Ziel R, Geiger T, Muller M, Fabbro D (1996) Inhibition of growth of human tumor cell lines in nude mice by an antisense oligonucleotide inhibitor of protein kinase C-α expression. Cancer Res 56:3499-3507

Defranco AL (1991) Immunosuppressants at work. Nature 352:754-755

De las Alas MM, Aebi S, Fink D, Howell SB, Los G (1997) Loss of DNA mismatch repair: effects on the rate of mutation to drug resistance. J Natl Cancer Inst 89:1537-1541

DeMaria R, Lenti L, Malisan F, d'Agostino F, Tomassini B, Zeuner A, Rippo MR, Testi R (1997) Requirement for GD3 ganglioside in CD95- and ceramide-induced apoptosis. Science 277:1652-1655

Denham DW, Franz MG, Denham W, Zervos EE, Gower WR Jr, Rosemurgy AS, Norman J (1998) Directed antisense therapy confirms the role of protein kinase C-alpha in the tumorigenicity of pancreatic cancer. Surgery 124:218-224

Denning MF, Wang Y, Nickoloff BJ, Wrone-Smith T (1998) Protein kinase cdelta is activated by caspase-dependent proteolysis during ultraviolet radiation-induced apoptosis of human keratinocytes. J Biol Chem 273:29995-30002

Dennis JU, Dean NM, Bennett CF, Griffith JW, Lang CM, Welch DR (1998) Human melanoma metastasis is inhibited following ex vivo treatment with an antisense oligonucleotide to protein kinase C-alpha. Cancer Lett 128:65-70

Desrivieres S, Volarevic S, Mercep L, Ferrari S (1997) Evidence for different mechanisms of growth inhibition of T-cell lymphoma by phrobol esters and concanavalin A. J Biol Chem 272:2470-2476

de Vente J, Kiley S, Garris T, Bryant W, Hooker J, Posekany K, Parker P, Cook P, Fletcher D, Ways DK (1995) Phorbol ester treatment of U937 cells with altered protein kinase C content and distribution induces cell death rather than differentiation. Cell Growth Differ 6:371-382

Diaconu CC, Szathmari M, Keri G, Venetianer A (1999) Apoptosis is induced in both drug-sensitive and multidrug-resistant hepatoma cells by somatostatin analogue TT-232. Br J Cancer 80:1197-1203

Diaz-Meco MT, Municio MM, Frutos S, Sanchez P, Lozano J, Sanz L, Moscat J (1996) The product of par 4, a gene induced during apoptosis, interacts selectively with the atypical isoforms of protein kinase C. Cell 86:777-786

Dietel M, Arps H, Lage H, Niendorf A (1990) Membrane vesicle formation due to acquired mitoxanthrone resistance in human gastric carcinoma cell line EPG85-257. Cancer Res 50:6100-6106

Dieter P, Fitzke E (1991) RO 31-8220 and RO 31-7549 show improved selectivity for PKC over staurosporine in macrophages. Biochem Biophys Res Commun 181:396-401

Dive C, Hickman, JA (1991) Drug-target interactions – only the first step in the commitment to a programmed cell-death. Br J Cancer 64:192-196

Doehmer J, Goeptar AR, Vermeulen NP (1993) Cytochromes P450 and drug resistance. Cytotechnology 12:357-366

Dolci ED, Abramson R, Xuan Y, Siegfried J, Yuenger KA, Yassa DS, Tritton TR (1993) Anomalous expression of P-glycoprotein in highly drug-resistant human KB cells. Int J. Cancer 54:302-308

Dolfini E, Dasdia T, Perletti G, Romagnoni M, Piccinini F (1993) Analysis of calcium-dependent protein kinase-C isoenzymes in intrinsically resistant cloned lines of LoVo cells: reversal of resistance by kinase inhibitor 1-(5-isoquinolinylsulfonyl) 2-methylpiperazine. Anticancer-Res 13:1123-1127

Dolfini E, Dasdia T, Arancia G, Molinari A, Calcabrini A, Scheper RJ, Flens MJ, Gariboldi MB, Monti E (1997) Characterization of a clonal human colon adenocarcinoma line intrinsically resistant to doxorubicin. Br J Cancer 76:67-76

Dong, Z, Ward NE, Fan D, Gupta, KP, O'Brian CA (1991) In vitro model for intrinsic drug resistance: effects of protein kinase C activators on chemosensitivity of cultured human colon cancer cells. Mol Pharmacol 39:563-569

Dooley NP, Baltuch GH, Groome N, Villemure JG, Yong VW (1998) Apoptosis is induced in glioma cells by antisense oligonucleotides to protein kinase C alpha and is enhanced by cycloheximide. Neuroreport 9:1727-1733

Doyle LA, Yang W, Abruzzo LV, Krogmann T, Gao Y, Rishi AK, Ross DG (1998) A multidrug resistance transporter from human MCF-7 breast cancer cells. Proc Natl Acad Sci USA 95:15665-15670

Drew L, Groome N, Hallam TJ, Warr JR, Rumsby MG (1994) Changes in protein kinase C subspecies protein expression and activity in a series of multidrug-resistant human KB carcinoma cell lines. Oncol Res 6:429-438

Drew L, Groome N, Warr JR, Rumsby MG (1996) Reduced daunomycin accumulation in drug-sensitive and multidrug-resistant human carcinoma KB cells following phorbol ester treatment: a potential role for protein kinase C in reducing drug influx. Oncol Res 8:249-257

Dwivedi RS, Wang LJ, Mirkin BL (1999) S-adenosylmethionine synthetase is overexpressed in murine neuroblastoma cells resistant to nucleoside analogue inhibitors of S-adenosylhomocysteine hydrolase: a novel mechanism of drug resistance. Cancer Res 56:1852-1856

Efferth T, Volm M (1992) Expression of protein kinase C in human renal cell carcinoma cells with inherent resistance to doxorubicin. Anticancer Res 12:2209-2211

Eguchi Y, Srinivasan A, Tomaselli KJ, Shimizu S, Tsujimoto Y (1999) ATP-dependent steps in apoptotic signal transduction. Cancer Res 59:2174-2181

Ek TP, Campbell MD, Deth RC, Gowraganahalli J (1989) Reduction of norepineph-rine-induced tonic contraction and phosphoinositide turnover in arteries of spontaneously hypertensive rats. A possible role for protein kinase C. Am J Hypertens 2:40-45

El-Deiry WS (1997) Role of oncogenes in resistance and killing by cancer therapeutic agents. Curr Opin Oncol 9:79-87

Emanuel SL, Chamberlin HA, Cohen D (1999) Antimitotic drugs cause increased tumorigenicity of multidrug resistant cells. Int J Oncol 14:487-494

Emoto Y, Manome Y, Meinhardt G, Kisaki H, Kharbanda S, Robertson M, Ghayur T, Wong WW, Kamen R, Weichselbaum R, Kufe D (1995) Proteolytic activation of protein kinase C delta by an ICE like protease in apoptotic cells. EMBO J 14:6148-6156

Endo K, Igarashi Y, Nisar M, Zhou QH, Hakomori S (1991) Cell membrane signaling as target in cancer therapy: inhibitory effect of N,N-dimethyl and N,N,N-trimethyl sphingosine derivatives on in vitro and in vivo growth of human tumor cells in nude mice. Cancer Res 51:1613-1618

Erickson LC (1991) The role of O-6 methylguanine DNA methyltransferase (MGMT) in drug resistance and strategies for its inhibition. Semin Cancer Biol 2:257-265

Evans CA, Lord JM, Owen-Lynch PJ, Johnson G, Dive C, Whetton AD (1995) Suppression of apoptosis by v-ABL protein tyrosine kinase is associated with nuclear translocation and activation of protein kinase C in an interleukin-3-dependent haemopoietic cell line. J. Cell Sci 108:2591-2598

Fabbro D, Regazzi R, Costa SD, Borner C, Eppenberger U (1986) Protein kinase C desensitization by phorbol esters and its impact on growth of human breast cancer cells. Biochem Biophys Res Commun 1986 135:65-73

Fabbro D, Buchdunger E, Wood J, Mestan J, Hofmann, F, Ferrari S, Mett H, O`Reilly T, Meyer T (1999) Inhibitors of protein kinases: CGP 41251, a protein kinase C inhibitor with potential as an anticancer agent. Pharmacol Ther 82:293-301

Fagerstrom S, Pahlman S, Gestblom C, Nanberg E (1996) Protein kinase C-epsilon is implicated in neurite outgrowth in differentiating human neuroblastoma cells. Cell Growth Differ 7:775-785

Fan D, Fidler IJ, Ward NE, Seid C, Earnest LE, Housey GM, O'Brian CA (1992) Stable expression of a cDNA encoding rat brain protein kinase C-beta I confers a multidrug-resistant phenotype on rat fibroblasts. Anticancer Res 12:661-667

Fan S, Chang JK, Smith ML, Duba D, Fornace AJ Jr, O´Connor PM (1997) Cells lacking CIP1/WAF1 genes exhibit preferential sensivity to cisplatin and nitrogen mustard. Oncogene 14:2127-2136

Ferraris C, Cooklis M, Polakowska RR, Haake AR (1997) Induction of apoptosis through the PKC pathway in cultured dermal papilla fibroblasts. Exp Cell Res 234: 37-46

Farrow SN, Brown R (1996) New members of the Bcl-2 family and their protein partners. Curr Opin Gen Develop 6:45-49

Ferry DR, Smith A, Malkhandi J, Fyfe DW, deTakats PG, Anderson D, Baker J, Kerr DJ (1996) Phase I clinical trial of the flavonoid quercetin: pharmacokinetics and evidence for in vivo tyrosine kinase inhibition. Clin Cancer Res 2:659-668

Findik D, Song Q, Hidaka H, Lavin M (1995) Protein kinase A inhibitors enhance radiation-induced apoptosis. J Cell Biochem 57:12-21

Fine RL, Patel J, Chabner BA (1988) Phorbol esters induce multidrug resistance in human breast cancer cells. Proc Natl Acad Sci USA, 85:582-586

Finkenzeller G, Marme D, Hug H (1992) Inducible overexpression of human protein kinase C in NIH 3T3 fibroblasts results in growth abnormalities. Cell Signall 4:163-177

Fisher DE (1994) Apoptosis in cancer therapy: crossing the threshold. Cell 78:539-542

Fitzsimmons SA, Workman P, Grever M, Paull K, Camalier R, Lewis AD (1996) Reductase enzyme expression across the National Cancer Institute tumor cell line panel: correlation with sensitivitry to mitomycin C and E09. J Natl Cancer Inst 88:259-269

Forbes IJ, Zalewski PD, Giannakis C, Cowled PA (1992) Induction of apoptosis in chronic lymphocytic leukemia cells and its prevention by phorbol ester. Exp Cell Res 198:367-72

Freemerman AJ, Turner AJ, Birrer MJ, Szabo E, Valerie K, Grant S (1996) Role of c-jun in human myeloid leukemia cell apoptosis induced by pharmacological inhibitors of protein kinase C. Mol Pharmacol 49:788-795

Freund A, Boos J, Harkin S, Schultze-Mosgau M, Veerman G, Peters GJ, Gescher A (1998) Augmentation of 1-beta-D-arabinofuranosylcytosine (Ara-C) cytotoxicity in leukaemia cells by co-administration with antisignalling drugs. Eur J Cancer 34:895-901

Frutos S, Moscat J, Diaz-Meco MT (1999) Cleavage of zetaPKC but not lamda/iota PKC by caspase-3 during UV-induced apoptosis. J Biol Chem 274:10765-10770

Fry DW, Jackson RC (1986) Membrane transport alterations as a mechanism of resistance to anticancer agents. Cancer Surv 5:47-79

Fyfe D, Ferry DR, Smith A, de Takats P, Baker J, Kerr DJ (1996) Phase I clinical trial of the tyrosine kinase inhibitor quercetin combined with carboplatin in advanced cancer. Ann Oncology 7: Suppl. 1, 263

Galski H, Lazarovici P, Gottesman MM, Murakata C, Matsuda Y, Hochman J (1995) KT-5720 reverses multidrug resistance in variant S49 mouse lymphoma cells transduced with the human MDR1 cDNA and in human multidrug-resistant carcinoma cells. Eur J Cancer 31A: 380-388

Garzotto M, White-Jones M, Jiang Y, Ehleiter D, Liao WC, Haimovitz-Friedman A, Fuks Z, Kolesnick R (1998) 12-O-tetradecanoylphorbol-13-acetate-induced apoptosis in LNCaP cells is mediated through ceramide synthase. Cancer Res 58:2260-2264

Geiger T, Muller M, Dean NM, Fabbro D (1998) Antitumor activity of a PKC-alpha antisense oligonucleotide in combination with standard chemotherapeutic agents against various human tumors transplanted into nude mice. Anticancer Drug Des 13:35-45

Geiges D, Marks F, Gschwendt M (1995) Loss of protein kinase Cδ from human HaCaT keratinocytes upon ras transfection is mediated by TGFα. Exp Cell Res 219:299-303

Geilen CC, Haase R, Buchner K, Wieder T, Hucho F, Reutter W (1991) The phospholipid analogue, hexadecylphosphocholine, inhibits protein kinase C in vitro and antagonises phorbol ester-stimulated cell proliferation. Eur J Cancer 27:1650-1653

Gekeler V, Boer R, Überall F, Ise W, Schubert C, Utz I, Hofmann J, Sanders KH, Schächtele C, Klemm K, Grunicke H (1996) Effects of the selective bisindolylmaleimide protein kinase C inhibitor GF 109203X on P-glycoprotein-mediated multidrug resistance. Br J Cancer 74:897-905

Germann UA, Chambers TC, Ambudkar SV, Licht T, Cardarelli CO, Pastan I, Gottesman MM (1996) Characterization of phosphorylation-defective mutants of human P-glycoprotein expressed in mammalian cells. J Biol Chem 217:1708-1716

Gescher A (1992) Towards selective pharmacological modulation of protein kinase C - opportunities for the development of novel antineoplastic agents. Br J Cancer 66:10-19

Gescher A (1998) Analogs of staurosporine: Potential anticancer drugs? Gen Pharmac 31:721-728

Ghayur T, Hugunin M, Talanian RV, Ratnofsky S, Quinlan C, Emoto Y, Pandey P, Datta R, Huang Y, Kharbanda S, Allen H, Kamen R, Wong W, Kufe D (1996) Proteolytic activation of protein kinase C delta by an ICE/CED 3 like protease induces characteristics of apoptosis. J Exp Med 184:2399-2404

Goekjian PG, Jirousek MR (1999) Protein kinase C in the treatment of disease: signal transduction pathways, inhibitors, and agents in development. Curr Med Chem 6:877-903

Goldie JH, Coldman AJ (1984) The genetic origin of drug resistance in neoplasms: implications for systemic therapy. Cancer Res 44:3643-3653

Gollapudi S, Soni V, Thadepalli H, Gupta S (1995) Role of protein kinase beta isozyme in multidrug resistance in murine leukemia P388/ADR cells. J Chemother 7:157-159

Goldstein DR, Cacace AM, Weinstein IB (1995) Overexpression of protein kinase C β1 in the SW480 colon cancer cell line causes growth suppression. Carcinogenesis 16:1121-1126

Gomez J, de la Hera A, Silva A, Pitton C, Garcia A, Rebello A (1994) Implication of protein kinase C in IL-2-mediated proliferation and apoptosis in a murine T cell clone. Exp Cell Res 213:178-182

Goodfellow HR, Sardini A, Ruetz S, Callaghan R, Gros P, McNaughton PA, Higgins CF (1996) Protein kinase C-mediated phosphorylation does not regulate drug transport by the human multidrug resistance P-glycoprotein. J Biol Chem 271:13668-13674

Gottesman MM, Pastan I (1993) Biochemistry of multidrug resistance mediated by the multidrug transporter. Annu Rev Biochem 62:385-427

Graham CH, Kobayashi H, Stankiewicz KS, Man S, Kapitain SJ, Kerbel RS (1994) Rapid acqisition of multicellular drug resistance after a single exposure of mammary tumor cells to antitumor aklylating agents. J Natl Cancer Inst 86:975-982

Grant S, Turner A, Bartimole TM, Nelms P, Joe VC, Jarvis WD (1994) Modulation of 1-β-D-arabinofuranosylcytosine-induced apoptosis in human promyelocytic leukemia cells by staurosporine and other inhibitors of protein kinase C. Oncol Res 6:87-99

Grant S, Jarvis DW (1996) Modulation of drug-induced apoptosis by interruption of protein kinase C signal transduction pathway: a new therapeutic strategy. Clin Cancer Res 2:1915-1920

Grant S, Turner AJ, Freemerman AJ, Wang Z, Kramer L, Jarvis WD (1996) Modulation of protein kinase C activity and calcium sensitive isoform expression in human myeloid leukemia cells by bryostatin 1: relationship to differentiation and ara C induced apoptosis. Exp Cell Res 228: 65-75

Grant S, Roberts J, Poplin E, Tombes MB, Kyle B, Welch D, Carr M, Bear HD (1998) Phase Ib trial of bryostatin 1 in patients with refractory malignancies. Clin Cancer Res 4:611-618

Gravitt KR, Ward NE, Fan D, Skibber JM, Levin B, O'Brian CA (1994) Evidence that protein kinase C-alpha activation is a critical event in phorbol ester-induced multiple drug resistance in human colon cancer cells. Biochem Pharmacol 48:375-381

Gray MO, Karliner JS, Mochly-Rosen D (1997) A selective epsilon-protein kinase C antagonist inhibits protection of cardiac myocytes from hypoxia-induced cell death. J Biol Chem 272:30945-30951

Griffiths G, Garrone B, Deacon E, Owen P, Pongracz J, Mead G, Bradwell A, Watters D, Lord J (1996) The polyether bistratene A activates protein kinase C-delta and induces growth arrest in HL60 cells. Biochem Biophys Res Commun 222:802-808

Gruber JR, Ohno S, and Niles RM (1992) Increased expression of protein kinase C alpha plays a key role in retinoic acid-induced melanoma differentiation. J Biol Chem 267:13356-13360

Gschwendt M, Furstenberger G, Rose-John S, Rogers M, Kittstein W, Pettit GR, Herald CL, Marks F (1988) Bryostatin 1, an activator of protein kinase C, mimics as well as inhibits biological effects of the phorbol ester TPA in vivo and in vitro. Carcinogenesis 9:555-562

Gschwendt M, Muller, HJ, Kielbassa K, Zang R, Kittstein W, Rincke G Marks F (1994) Rottlerin, a novel protein kinase inhibitor. Biochem Biophys Res Commun 199: 93-98

Gschwendt M (1999) Protein kinase Cδ. Eur J Biochem 259:555-564

Guillem JG, O'Brian CA, Fitzer CJ, Forde KA, LoGerfo P, Treat M, Weinstein IB (1987) Altered levels of protein kinase C and Ca²⁺-dependent protein kinases in human colon carcinomas. Cancer Res 47:2036-2039

Gupta S, Patel K, Singh H, Gollapudi S (1994) Effect of Calphostin C (PKC inhibitor) on daunorubicin resistance in P388/ADR and HL60/AR cells: reversal of drug resistance possibly via P-glycoprotein. Cancer Lett 76:139-145

Gupta KP, Ward NE, Gravitt KR, Bergman PJ, O'Brian CA (1996) Partial reversal of multidrug resistance in human breast cancer cells by an N-myristoylated protein kinase C-alpha pseudosubstrate peptide. J Biol Chem 271:2102-2111

Haggerty HG, Monroe JG (1994) A mutant of the WEHI-231 B lymphocyte line that is resistant to phorbol esters is still sensitive to antigen receptor-mediated growth inhibition. Cell Immunol 154:166-180

Hagiwara M, Hachiya T, Watanabe M, Usuda N, Iida F, Tamai K, Hidaka H (1990) Assessment of protein kinase C isoenzymes by immunoassay and overexpression of type II in thyroid adenocarcinoma. Cancer Res 50:5515-5519

Hähnel R, Gschwendt M (1995) The interaction between protein kinase C (PKC) and estrogens. Int J Oncol 7:11-16

Haimovitz-Friedman A, Kann CC, Ehleiter D, Persaud RS, McLoughlin M, Fuks Z, Kolesnick RN (1994a) Ionizing radiation acts on cellular membranes to generate ceramide and initiate apoptosis. J Exp Med 180:525-535

Haimovitz-Friedman A, Balaban N, McLoughlin M, Ehleiter D, Michaeli J, Vlodavsky I, Fuks Z (1994b) Protein kinase C mediates basic fibroblast growth factor protection of endothelial cells against radiation-induced apoptosis. Cancer Res 54: 2591-2597

Hait WN, DeRosa WT (1991) The role of the phorbol ester receptor/protein kinase C in the sensitivity of leukemic cells to anthracyclines. Cancer Commun 3:77-81

Hallahan DE, Virudachalam S, Schwartz JL, Panje N, Mustafi R, Weichselbaum RR (1992) Inhibition of protein kinases sensitizes human tumor cells to ionizing radiation. Radiat Res 129:345-350

Han EK, Begemann M, Sgambato A, Soh JW, Doki Y, Xing WQ, Liu W, Weinstein IB (1996) Increased expression of cyclin D1 in a murine mammary epithelial cell line induces p27kip1, inhibits growth, and enhances apoptosis. Cell Growth Differ 7:699-710

Hannun YA (1998) Functions of ceramide in coordinating cellular responses to stress. Science 274:1855-1859

Hardy SP, Goodfellow HR, Valverde MA, Gill DR, Sepulveda V, Higgins CF (1995) Protein kinase C-mediated phosphorylation of the human multidrug resistance P-glycoprotein regulates cell volume-activated chloride channels EMBO J 14:68-75

Harkin ST, Cohen GM, Gescher A (1998) Modulation of apoptosis in rat thymocytes by analogs of staurosporine: lack of direct association with inhibition of protein kinase C. Mol Pharmacol 54:663-670

Hatada S, Sakanoue Y, Kusunoki M, Kobayashi A, Utsunomiya J (1992) Protein kinase C activity in human thyroid carcinoma and adenoma. Cancer 70:2918-2922

Haq R, Zanke B (1998) Inhibition of apoptotic signaling pathways in cancer cells as a mechanism of chemotherapy resistance. Cancer Metastasis Rev 17:233-239

Heesbeen EC, Verdonck LF, Staal GE, Rijksen, G (1994) Protein kinase C is not involved in the cytotoxic action of 1-octadecyl-2-O-methyl-sn-glycerol-3-phosphocholine in HL-60 and K562 cells. Biochem Pharmacol 47:1481-1488

Hilgard P, Klenner T, Stekar J, Unger C (1993) Alkylphosphocholines: a new class of membrane-active anticancer agents. Cancer Chemother Pharmacol 32:90-95

Hirata J, Kikuchi Y, Kita T, Imaizumi E, Tode T, Ishii K, Kudoh K, Nagata I (1993) Modulation of sensitivity of human ovarian cancer cells to cis-diamminedichloroplatinum(II) by 12-O-tetradecanoylphorbol-13-acetate and D,L-buthionine-S,R-sulphoximine. Int J Cancer 55:521-527

Höckel M, Schlenger K, Aral B, Mitze M, Schaffer U, Vaupel P (1996) Association between tumor hypoxia and malignant progression in advanced cancer of the uterine cervix. Cancer Res 56:4509-4515

Hoelting T, Duh QY, Clark OH, Herfarth C (1996) Tamoxifen antagonizes proliferation and invasion of estrogen receptor-negative metastatic follicular thyroid cancer cells via protein kinase C. Cancer Lett 100: 89-93

Hocevar BA, Fields AP (1991) Selective translocation of βII-PKC to the nucleus of human promyelocytic (HL-60) leukemia cells. J Biol Chem 266:28-33

Hofmann J, Doppler W, Jakob A, Maly K, Posch L, Überall F, Grunicke H (1988) Enhancement of the antiproliferative effect of cis diamminedichloroplatinum(II) and nitrogen mustard by inhibitors of protein kinase C. Int J Cancer 42:382-388

Hofmann J, Überall F, Posch L, Maly K, Herrmann DBJ, Grunicke H (1989) Synergistic enhancement of the anti proliferative activity of cis diamminedichloroplatinum(II) by the new ether lipid analogue BM 41 440, an inhibitor of protein kinase C. Lipids 24:312-317

Hofmann J, Fiebig HH, Winterhalter BR, Berger DP, Grunicke H. (1990) Enhancement of the antiproliferative activity of cis diamminedichloroplatinum(II) by quercetin. Int J Cancer 45:536-539

Hofmann J, Gekeler V, Ise W, Noller A, Mitterdorfer J, Hofer S, Utz I, Gotwald M, Boer R, Glossmann H, Grunicke H (1995) Mechanism of action of dexniguldipine-HCl (B8509-035), a new potent modulator of multidrug resistance. Biochem Pharmacol 49:603-609

Hofmann J, Utz I, Spitaler M, Hofer S, Rybczynska M, Beck WT, Herrmann DB, Grunicke H (1997) Resistance to the new anti-cancer phospholipid ilmofosine (BM 41 440). Br J Cancer 76:862-869

Hofmann J (1997) The potential for isoenzyme-selective modulation of protein kinase C. FASEB J 11:649-669

Hofmann K, Dixit VM (1999) Reply to Kolesnick and Hannun, and Perry and Hannun. Trends Biochem Sci 282, 227

Hooijberg JH, Broxterman HJ, Kool M, Assaraf YG, Peters GJ, Noordhuis P, Scheper RJ, Borst P, Pinedo HM, Jansen G (1999) Antifolate resistance mediated by multidrug resistance proteins MRP1 and MRP2. Cancer Res 59:2532-2535

Hornung RL, Pearson JW, Beckwith M, Longo DL (1992) Preclinical evaluation of bryostatin as an anticancer agent against several murine tumor cell lines: in vitro versus in vivo activity. Cancer Res 52:101-107

Housey GM, Johnson MD, Hsiao WL, O'Brian CA, Murphy JP, Kirschmeier P, Weinstein IB (1988) Overproduction of protein kinase C causes disordered growth control in rat fibroblasts. Cell 52:343-354

Hsiao WL, Housey GM, Johnson MD, Weinstein IB (1989) Cells that overproduce protein kinase C are more susceptible to transformation by an activated. H-ras oncogene. Mol Cell Biol 9:2641-2647

Hu H (1996) Recent discovery and development of selective protein kinase C inhibitors. Drug Discov Today 1:438-447

Hu YP, Robert J (1997) Inhibition of protein kinase C in multidrug-resistant cells by modulators of multidrug resistance. J Cancer Res Clin Oncol 123:201-210

Huang JC, Zamble DB, Reardon JT, Lippard SJ, Sancar A (1994) HMG-domain proteins specifically inhibit the repair of the major DNA adduct of the anticancerdrug cisplatin by human excision nuclease. Proc Natl Acad Sci USA 91:10394-10398

Hug H, Sarre TF (1993) Protein kinase C isoenzymes: divergence in signal transduction? Biochem J 291:329-343

Hunakova L, Sulikova M, Duraj J, Sedlak J, Chorvath B (1996) Stimulation of 1-(beta-D-arabinofuranosyl)cytosine (AraC)-induced apoptosis in the multidrug resistant human promyelocytic leukemia cell lines with protein kinase inhibitors. Neoplasma 43:291-295

Hundle B, McMahon T, Dadgar J, Messing RO (1995) Overexpression of epsilonprotein kinase C enhances nerve growth factor-induced phosphorylation of mitogen-activated protein kinases and neurite outgrowth. J Biol Chem 270:30134-30140

Husain A, Yan XJ, Rosales N, Aghajanian C, Schwartz GK, Spriggs DR (1997) UCN-01 in Ovary Cancer Cells: Effective as a Single Agent and in Combination with cis-Diamminedichloroplatinum(II)Independent of p53 Status. Clin Cancer Res 3:2089-2097

Husain A, He G, Venkatraman ES, Spriggs DR (1998) BRCA1 up-regulation is associated with repair-mediated resistance to cis-diamminedichloroplatinum(II). Cancer Res 58:1120-1123

Ikegami Y, Yano S, Nakao K, Fujita F, Fujita M, Sakamoto Y, Murata N, Isowa K (1995) Antitumour activity of the new selective protein kinase C inhibitor 4'-N-benzoyl staurosporine on murine and human tumour models. Arzneim Forsch 45:1225-1230

Ikegami Y, Yano S, Nakao K (1996a) Antitumour effect of CGP 41251, a new selective protein kinase C inhibitor on human non-small cell lung cancer cells. Jpn J Pharmac 70:65-72

Ikegami Y, Yano S, Nakao K (1996b) Effects of the new selective kinase C inhibitor 4'-N-benzoyl staurosporine on cell cycle distribution and growth inhibition in human small cell lung cancer cells. Arzneim.-Forsch 46:201-204

Ilic D, Almeida EAC, Shlaepfer DD, Dazin P, Aizawa S, Damsky CH (1998) Extracellular matrix survival signals transduced by focal adhesion kinase suppress p53-mediated apoptosis. J Cell Biol 143:547-560

Imbra RJ, Karin M (1987) Metallothionein gene expression is regulated by serum factors and activators of protein kinase C. Mol Cell Biol 7:1358-1363

Inoguchi T, Battan R, Handler E, Sportsman JR, Heath W, King GL (1992) Preferential elevation of protein kinase C isoform βII and diacylglycerol levels in the aorta and heart of diabetic rats: differential reversibility to glycemic control by islet cell transplantation. Proc Natl Acad Sci USA 89:11059-11063

Iseki R, Mukai M, Iwata M (1991) Regulation of T lymphocyte apoptosis. Signals for the antagonism between activation- and glucocorticoid-induced death. J Immunol 147:4286-4292

Ishii HH, Gobe GC (1993) Epstein-Barr virus infection is associated with increased apoptosis in untreated and phorbol ester-treated human Burkitt's lymphoma (AW-Ramos) cells. Biochem Biophys Res Commun 192:1415-1423

Ishii H, Jirousek MR, Koya D, Takagi C, Xia P, Clermont A, Bursell S-E, Kern TS, Ballas LM, Heath WF, Stramm LE, Feener EP, King GL (1996) Amelioration of vascular dysfunctions in diabetic rats by an oral PKCβ inhibitor. Science 272:728-731

Isonishi S, Andrews PA, Howell SB (1990) Increased sensitivity to cis diamminedichloroplatinum(II) in human ovarian carcinoma cells in response to treatment with 12-O-tetradecanoylphorbol-13-acetate. J Biol Chem 265:3623-3627

Isonishi S, Hom DK, Thiebaut FB, Mann SC, Andrews PA, Basu A, Lazo JS, Eastman A, Howell SB (1991) Expression of the c-Ha-ras oncogene in mouse NIH 3T3 cells induces resistance to cisplatin. Cancer Res 51:5903-5909

Issandou M, Bayard F, Darbon JM (1988) Inhibition of MCF-7 cell growth by 12-O-tetradecanoylphorbol-13-acetate and 1,2-dioctanoyl-sn-glycerol: distinct effects on protein kinase C activity. Cancer Res 48:6943-6950

Issandou M, Darbon JM (1988) 1,2-Dioctanoyl-glycerol induces a discrete but transient translocation of protein kinase C as well as the inhibition of MCF-7 cell proliferation. Biochem Biophys Res Commun 151:458-465

Iwata M, Iseki R, Sato K, Tozawa Y, Ohoka Y (1994) Involvement of protein kinase C epsilon in glucocorticoid induced apoptosis in thymocytes. Int Immunol 6:431-438

Jamieson L, Carpenter L, Biden TJ, Fields AP (1999) Protein kinase Cι activity is necessary for Bcr-Abl-mediated resistance to drug-induced apoptosis. J Biol Chem 274:3927-3930

Jarvis WD, Turner AJ, Povirk LF, Traylor RS, Grant S (1994a) Induction of apoptotic DNA fragmentation and cell death in HL 60 human promyelocytic leukemia cells by pharmacological inhibitors of protein kinase C. Cancer Res 54:1707-1714

Jarvis WD, Povirk LF, Turner AJ, Traylor RS, Gewirtz DA, Pettit GR, Grant S (1994b) Effects of bryostatin 1 and other pharmacological activators of protein kinase C on 1 [beta D arabinofuranosyl]cytosine induced apoptosis in HL 60 human promyelocytic leukemia cells. Biochem Pharmacol 47:839-852

Jarvis WD, Fornari FA Jr, Tombes RM, Erukulla RK, Bittman R, Schwartz GK, Dent P, Grant S (1998) Evidence for involvement of mitogen-activated protein kinase, rather than stress-activated protein kinase, in potentiation of 1-beta-D-arabinofuranosylcytosine-induced apoptosis by interruption of protein kinase C signaling. Mol Pharmacol 54:844-856

Jayson GC, Crowther D, Prendiville J, McGown AT, Scheid C, Stern P, Young R, Brenchley P, Chang J, Owens S, Pettit GR (1995) A phase I trial of bryostatin 1 in patients with advanced malignancy using a 24 hour intravenous infusion. Br J Cancer 72:461-468

Jiang JB, Johnson MG, Defauw JM, Beine TM, Ballas LM, Janzen WP, Loomis CR, Seldin J, Cofield D, Adams L, Cianciolo G, Degen D, Vonhoff DD (1992) Novel non-cross resistant diaminoanthraquinones as potential chemotherapeutic agents. J Med Chem 35:4259-4263

Jin L, Maeda T, Chandler WF, Lloyd RV (1993) Protein kinase C (PKC) activity and PKC messenger RNAs in human pituitary adenomas. Am J Pathol 142:569-578

Johnstone RW, Cretney E, Smyth MJ (1999) P-glycoprotein protects leukemia cells against caspase-dependent, but not caspase-independent, cell death. Blood 93:1075-1085

Juliano RL, Ling V (1976) A surface glycoprotein modulating drug permeability in Chinese hamster ovary cell mutants. Biochim. Biophys. Acta 455:152-162

Kahl-Rainer P, Karner-Hanusch J, Weiss W, Marian B (1994) Five of six protein kinase C isoenzymes present in normal mucosa show reduced protein levels during tumor development in the human colon. Carcinogenesis 15:779-782

Kaina B, Lohrer H, Karin M, Herrlich P (1990) Overexpressed human metallothionein II gene protects Chinese hamster ovary cells from killing by alkylating agents. Proc Natl Acad Sci USA 87:2710-2714

Kakawami K, Futami H, Takahara J, Yamaguchi K (1996) UCN-01, 7-hydroxylstaurosporine, inhibits kinase activity of cyclin-dependent kinases and reduces phosphorylation of the retinoblastoma susceptibility gene product in A549 human lung cancer cell line. Biochem Biophys Res Commun 219:778-783

Kampfer S, Hellbert K, Villunger A, Doppler W, Baier G, Grunicke HH, Überall F (1998) Transcriptional activation of c-fos by oncogenic Ha-Ras in mouse mammary epithelial cells requires the combined activities of PKC-λ, ε and ζ. EMBO J 17:4046-4055

Kanter P, Leister KJ, Tomei LD, Wenner PA, Wenner CE (1984) Epidermal growth factor and tumor promoters prevent DNA fragmentation by different mechanisms. Biochem Biophys Res Commun 118:392-399

Kariya K-I, Kawahara Y, Tsuda T (1987) Possible involvement of protein kinase C in platelet-derived growth factor-stimulated DNA synthesis in vascular smooth muscle cells. Atherosclerosis 83:251-255

Keenan C, Thompson S, Knox K, Pears C (1999) Protein kinase C-alpha is essential for Ramos-BL B cell survival. Cell Immunol 196:104-109

Kelley SL, Basu A, Teicher BA, Hacker MP, Hamer DH, Lazo JS (1988) Overexxpression of metallothionein confers resistance to anticancer drugs. Science 241:1815-1818

Kelly ML, Tang Y, Rosensweig N, Clejan S, Beckman BS (1998) Granulocyte-macrophage colony-stimulating factor rescues TF-1 leukemia cells from ionizing radiation-induced apoptosis through a pathway mediated by protein kinase Cα. Blood 92:416-424

Kennedy MJ, Prestigiacomo LJ, Tyler G, May WS, Davidson NE (1992) Differential effects of bryostatin 1 and phorbol ester on human breast cancer cell lines. Cancer Res 52:1278-1283

Kharbanda S, Datta R, Kufe D (1991) Regulation of c jun gene expression in HL 60 leukemia cells by 1 beta D arabinofuranosylcytosine. Potential involvement of a protein kinase C dependent mechanism. Biochemistry 30:7947-7952

Kieser A, Seitz T, Adler HS, Coffer P, Kremmer E, Crespo P, Gutkind JS, Henderson DW, Mushinski JF, Kolch W, Mischak H (1996) Protein kinase C-zeta reverts v-raf transformation of NIH-3T3 cells. Genes Dev 10:1455-1466

Kiley SC, Clark KJ, Duddy SK, Welch DR, Jaken S (1999) Increased protein kinase Cδ in mammary tumor cells: relationship to transformation and metastatic progression. Oncogene 18:6748-6757

Killion JJ, Beltran P, O'Brian CA, Yoon SS, Fan D, Wilson MR, Fidler IJ (1995) The antitumor activity of doxorubicin aganist drug-resistant murine carcinom is enhanced by oral administration of a synthetic staurosporine analogue, CGP 41251. Oncol Res 7:453-459

Kim CY, Giaccia AJ, Strulovici B, Brown JM (1992) Differential expression of protein kinase C epsilon protein in lung cancer cell lines by ionising radiation. Br J Cancer 66:844-849

Kim CH, Gollapudi S, Lee T, Gupta S (1997) Altered expression of the genes regulating apoptosis in multidrug resistant human myeloid leukemia cell lines overexpressing MDR1 or MRP gene. Int J Oncol 11:945-950

Kinsella AR, Smith D, Pickard M (1997) Resistance to chemotherapeutic antime-
tabolites: a function of salvage pathway involvement and cellular response to
DNA damage. Br J Cancer 75:935-945

Kinter AL, Poli G, Maury W, Folks TM, Fauci AS (1990) Direct and cytokine-
mediated activation of protein kinase C induces human immunodeficiency virus
expression in chronically infected promonocytic cells. J Virol 64:4306-4312

Kiss Z, Deli, E, Vogler WR, Kuo JF (1987) Antileukemic agent alkyllysophospholipid
regulates phosphorylation of distinct proteins in HL60 and K562 cells and differentiation of HL60 cells promoted by phorbol ester. Biochem Biophys Res Commun 142:661-666

Kizaki H, Tadakuma T, Odaka C, Muramatsu J, Ishimura Y (1989) Activation of a
suicide process of thymocytes through DNA fragmentation by calcium iono-
phores and phorbol esters. J Immunol 143:1790-1794

Klenner T, Engel J, Hilgard P (1998) Alkylphosphocholines: An update. Drugs of
Today 34:Suppl. F, Prous Science 1-2

Knox KA, Johnson GD, Gordon J (1993) A study of protein kinase C isozyme distribution in relation to Bcl-2 expression during apoptosis of epithelial cells in vivo.
Exp Cell Res 207:68-73

Knox KA, Gordon J (1994) Protein tyrosine kinases couple the surface immunoglobulin of germinal center B cells to phosphatidylinositol-dependent and
-independent pathways of rescue from apoptosis. Cell Immunol 155:62-76

Kobayashi H, Man S, Graham CH, Kapitain SJ, Teicher BA, Kerbel R.S (1993) Acquired multicellular-mediated resistance to alkylating agents in cancer. Proc Natl
Acad Sci. USA 90:3294-3298

Kobayashi D, Watanabe N, Yamauchi N, Tsuji N, Sato T, Sasaki H, Okamoto T,
Niitsu Y (1997) Protein kinase C inhibitors augment tumor-necrosis-factor-
induced apoptosis in normal human diploid cells. Chemotherapy 43:415-423

Kolch W, Heidecker B, Kochs G, Hummel R, Vahidi M, Mischak H, Finkenzeller G,
Marme D, Rapp UR (1993) Protein kinase C alpha activates RAF-1 by direct
phosphorylation. Nature 364:249-252

Kolesnick R, Hannun YA (1999) Ceramide and apoptosis. Trends Biochem Sci
282:224-225

Kondo S, Barnett GH, Hara H, Morimura T, Takeuchi J (1995a) MDM2 protein confers the resistance of a human glioblastoma cell line to cisplatin-induced apoptosis. Oncogene 10:2001-2006

Kondo Y, Woo ES, Michalska AE, Choo AKH, Lazo JS (1995b) Metallothionein null
cells have increased sensitivity to anticancer drugs. Cancer Res 55:2021-2023

Kool M, van der Linden M, de Haas M, Scheffer GL, de Vree JM, Smith AJ, Jansen G,
Peters GJ, Ponne N, Scheper RJ, Elferink RP, Baas F, Borst P (1999) MRP3, an or-
ganic anion transporter able to transport anti-cancer drugs. Proc Natl Acad Sci
USA 96:6914-6919

Kopnin BP, Stromskaya TP, Kondratov RV, Ossovskaya S, Pugacheva EN, Rybalkina
EY, Khokhlova OA, Chumakov PM (1995) Influence of exogenous ras and p53 on
P-glycoprotein function in immortalized rodent fibroblasts. Oncol Res 7:299-306

Kopp R, Noelke B, Sauter G, Schildberg FW, Paumgartner G, Pfeiffer A (1991) Altered protein kinase C activity in biopsies of human colonic adenomas and carcinomas. Cancer Res 51:205-210

Koriyma H, Kouchi Z, Umeda T, Saido TC, Momoi T, Ishiura S, Suzuki K (1999) Proteolytic activation of protein kinase C δ and ε by caspase-3 in U937 cells during chemotherapeutic agent-induced apoptosis. Cell Signal 11:831-838

Koseki C, Herzlinger D, al-Awqati Q (1992) Apoptosis in metanephric development. J Cell Biol 119:1327-1333

Kraft AS Reeves JA, Askendel CL (1988) Differing modulation of PKC by bryostatin 1 and phorbol esters in JB6 mous epidermal cells. J Biol Chem 263:8437-8442

Kusunoki M, Sakanoue Y, Hatada T, Yanagi H, Yamamura T, Utsunomiya J (1992) Protein kinase C activity in human colonic adenoma and colorectal carcinoma. Cancer 69: 24-30

Kwiatkowska-Patzer B, Domanska-Janik K (1991) Increased 19 kDa protein phosphorylation and protein kinase C activity in pressure-overload cardiac hypertrophy. Basic Res Cardiol 86:402-409

Lallena MJ, Diaz-Meco MT, Bren G, Paya CV, Moscat J (1999) Activation of IkappaB kinase beta by protein kinase C isoforms. Mol Cell Biol 19:2180-2188

La Porta CA, Tessitore L, Comolli R (1997) Changes in protein kinase C alpha, delta and in nuclear beta isoform expression in tumour and lung metastatic nodules induced by diethylnitrosamine in the rat. Carcinogenesis 4:715-719

La Porta CAM, Dolfini E, Comolli R (1998) Inhibition of protein kinase C-α isoform enhances the P-glycoprotein expression and the survival of LoVo human colon adenocarcinoma cellst to doxorubicin exposure. Br J Cancer 78:1283-1287

Laouar A, Glesne D, Huberman E (1999) Involvement of protein kinase C-beta and ceramide in tumor necrosis factor-alpha-induced but not Fas-induced apoptosis of human myeloid leukemia cells. J Biol Chem 274:23526-23534

Laredo J, Huynh A, Muller C, Jaffrezou JP, Bailly JD, Cassar G, Laurent G, Demur C (1994) Effect of the protein kinase C inhibitor staurosporine on chemosensitivity to daunorubicin of normal and leukemic fresh myeloid cells. Blood 84:229-237

Larocca LM, Teofili L, Sica S, Piantelli M, Maggiano N, Leone G, Ranelletti FO (1995) Quercetin inhibits the growth of leukemic progenitors and induces the expression of transforming growth factor-beta 1 in these cells. Blood 85:3654-3661

Lavie Y, Cao HAT, Volner A, Lucci A, Han TY, Geffen V, Giuliano AE, Cabot MC (1997) Agents that reverse multidrug resistance, tamoxifen, verapamil, and cyclosporin A, block glycosphingolipid metabolism by inhibiting ceramide glycosylation in human cancer cells. J Biol Chem 272:1682-1687

Lavie Y, Zhang ZC, Cao H-T, Han T-Y, Johnes RC, Liu YY, Jarman M, Hardcastle IR, Giuliano AE, Cabot MC (1998) Tamoxifen induces selective membrane association of protein kinase C epsilon in MCF-7 human breast cancer cells. Int J Cancer 77:928-932

Lavin MF, Watters D, Song Q (1996) Role of protein kinase activity in apoptosis. Experientia 52:979-994

Leach KL, James ML, Blumberg PM (1983) Characterization of a specific phorbol ester aporeceptor in moujse brain cytosol. Proc Natl Acad Sci USA 80:4208-4212

Lee SA, Karaszkiewicz JW, Anderson WB (1992) Elevated level of nuclear protein kinase C in multidrug-resistant MCF-7 human breast carcinoma cells. Cancer Res 52:3750-3759

Lee JY, Hannun YA, Obeid LM (1996a) Ceramide inactivates cellular protein kinase C alpha. J Biol Chem 271:13169-13174

Lee HW, Smith L, Pettit GR, Bingham-Smith J (1996b) Dephosphorylation of acti-
    vated protein kinase C contributes to downregulation by bryostatin. Am J
    Physiol 271(1 Pt 1): C304-311
Leitges M, Schmedt C, Guinamard R, Davoust J, Schaal S, Stabel S, Tarakhovsky A
    (1996) Immunodeficiency in protein kinase Cbeta-deficient mice. Science
    273:788-791
Leszczynski D, Joenvaara S, Foegh ML (1995) Apoptosis of rat vascular smooth
    muscle cells is regulated by δ and ζ but not by α and ε isozymes of protein kinase
    C. Mol Biol Cell 6, Suppl:246a
Leszczynski D, Joenvaara S, Foegh ML (1996) Protein kinase C alpha regulates pro-
    liferation but not apoptosis in rat coronary vascular smooth muscle cells. Life Sci
    58:599-606
Levin DE, Fields FO, Kunisawa R, Bishop JM, Thorner J (1990) A candidate protein
    kinase C gene, PKC1, is required for the S. cerevisiae cell cycle. Cell 62:213-224
Levin D, Bartrett-Heubusch E (1992) Mutants in the S cerevisiae PKC1 gene display a
    cell cycle-specific osmotic stability defect. J Cell Biol 116:1079-1088
Levine AJ (1997) p53, the cellular gatekeeper for growth and division.Cell 88:323-331
Levy MF, Pocsidio J, Guillem JG, Forde K, LoGerfo P, Weinstein IB (1993) Decreased
    levels of protein kinase C enzyme activity and protein kinase C mRNA in pri-
    mary colon tumors. Dis Colon Rectum 36:913-921
Li H, Zhao L, Yang Z, Funder JW, Liu JP (1998) Telomerase is controlled by protein
    kinase Calpha in human breast cancer cells. J Biol Chem 273:33436-33442
Liu YY, Han TY, Giuliano AE, Cabo MC (1999) Expression of glucosylceramide
    synthase, converting ceramide to glucosylceramide, confers adriamycin resis-
    tance in human breast cancer cells. J Biol Chem 274:1140-1146
Livneh E, Fishman DD (1997) Linking protein kinase C to cell-cycle control. Eur J
    Biochem 248:1-9
Lotem J, Cragoe EJ Jr, Sachs L (1991) Rescue from programmed cell death in leuke-
    mic and normal myeloid cells. Blood 78:953-960
Lowe SW, Ruley HE, Jacks T, Housman DE (1993) p53-dependent apoptosis modu-
    lates the cytotoxicity of anticancer agents. Cell 74:957-967
Lozano J, Berra E, Municio MM, Diaz-Meco MT, Dominguez I, Sanz L, Moscat J
    (1994) Protein kinase C zeta isoform is critical for kappa B-dependent promoter
    activation by sphingomyelinase. J Biol Chem 269:19200-19202
Lucas M, Sanchez-Margalet V, Sanz A, Solano F (1994) Protein kinase C activation
    promotes cell survival in mature lymphocytes prone to apoptosis. Biochem
    Pharmacol 47:667-672
Lucas M, Sanchez Margalet V (1995) Protein kinase C involvement in apoptosis. Gen
    Pharmacol 26:881-887
Lush RM, Dsenderowicz A, Figg WD, Headlee D, Inoue K, Sausville EA (1997) Sur-
    prising pharmacokinetics of UCN-01 in patients with refractory neoplasms may
    be due to high degree of plasma protein binding. Proc Am Assoc Cancer Res
    38:600
Ma LD, Marquardt D, Takemoto L., Center MS (1991) Analysis of P-glycoprotein
    phosphorylation in HL60 cells isolated for resistance to vincristine. J Biol Chem
    266:5593-5599
Ma J, Maliepaard M, Kolker HJ, Verweij J, Schellens JHM (1998a) Abrogated energy-
    dependent uptake of cisplatin in a cisplatin-resistant subline of the human ovar-
    ian cancer cell line IGROV-1. Cancer Chemother Pharmacol 41:186-192

Ma J, Maliepard M, Nooter K, Loos WJ, Kolker HJ, Verweij J, Stoter G, Schellens JHM (1998b) Reduced cellular accumulation of topotecan: a novel mechanism of resistance in a human ovarian cancer cell line. Br J Cancer 77:1645-1652

MacFarlane DE, O'Donnell PS (1993) Phorbol ester induces apoptosis in HL-60 promyelocytic leukemia cells but not in HL-60 PET mutant. Leukemia 7:1846-1851

MacFarlane DE, Manzel L (1994) Activation of beta-isozyme of protein kinase C (PKC beta) is necessary and sufficient for phorbol ester-induced differentiation of HL-60 promyelocytes. Studies with PKC beta-defective PET mutant. J Biol Chem 269:4327-4331

Magnelli L, Chiarugi V (1997) Regulation of p53 by protein kinase C during multistage carcinogenesis. J. Cancer Res Clin Oncol 123:365-369

Mailhos C, Howard MK, Latchman DS (1994) A common pathway mediates retinoic acid and TPA dependent programmed cell death (apoptosis) of neuronal cells. Brain Res 644:7-12

Manni A, Buckwalter E, Etindi R, Kunselman S, Rossini A, Mauger D, Dabbs D, Demers L (1996) Induction of a less aggressive breast cancer phenotype by protein kinase C-α and -β overexpression. Cell Growth Diff 7:1187-1198

Mansat V, Laurent G, Levade T, Bettaieb A, Jaffrezou JP (1997) The protein kinase C activators phorbol esters and phosphatidylserine inhibit neutral sphingomyelinase activation, ceramide generation, and apoptosis triggered by daunorubicin. Cancer Res 57:5300-5304

Marte B, Meyer T, Stabel S, Standke GJ, Jaken S, Fabbro D, Hynes NE (1994) Protein kinase C and mammary cell differentiation: involvement of protein kinase C alpha in the induction of beta-casein expression. Cell Grwoth Differ 5:239-247

Martelli AM, Sang N, Borgatti P, Capitani S, Neri LM (1999) Multiple biological responses activated by nuclear protein kinase C. J Cell Biochem 74:499-521

Martinez-Zaguilan R, Raghunand N, Lynch RM, Bellamy W, Martinez GM, Rojas B, Smith D, Dalton WS, Gillies RJ (1999) pH and drug resistance. I. Functional expression of plasmalemmal V-type H^+-ATPase in drug-resistant human breast carcinoma cell lines. Biochem Pharmacol 57:1037-1046

Martiny-Baron G, Kazanietz MG, Mischak H, Blumberg PM, Kochs G, Hug H, Marme D, Schachtele C (1993) Selective inhibition of protein kinase C isozymes by the indolocarbazole Go 6976. J Biol Chem 268:9194-9197

Masanek U, Stammler G, Volm M (1997) Messenger RNA expression of resistance proteins and related factors in human ovarian carcinoma cell lines resistant to doxorubicin, taxol and cisplatin. Anticancer Drugs 8: 189-198

Masumoto N, Nakano S, Jujishima H, Hohno K, Niho Y (1999) v-src induces cisplatin resistance by increasing the repair of cisplatin-DNA interstrand crosslinks in human gallbladder adenocarcinoma cells. Int J Cancer 80:731-737

Mathias S, Pena LA, Kolesnick RN (1998) Signal transduction of stress via ceramide. Biochem J 335 (Pt 3):465-480

Matsumoto T, Tani E, Yamaura I, Miyaji K, Kaba K (1995) Effects of protein kinase C modulators on multidrug resistance in human glioma cells. Neurosurgery 36:565-571

Mattern J, Eichhorn U, Kaina B, Volm M (1998) O^6-Methylguanine-DNA methyltransferase activity and sensitivity to cyclophoshamide and cisplatin in human lung tumor xenografts. Int. J Cancer 77:919-922

Mayne GC, Murray AW (1998) Evidence that protein kinase Cε mediates phorbol ester inhibition of calphostin C- and tumor necrosis factor-α-induced apoptosis in U937 histiocytic lymphoma cells J Biol Chem 273:24115-24121

McBain JA, Pettit GR, Mueller GC (1990) Phorbol esters activate proteoglycan metabolism in human colon cancer cells en route to terminal differentiation. Cell Growth Differ 1:281-291

McConkey DJ, Hartzell P, Jondal M, Orrenius S (1989) Inhibition of DNA fragmentation in thymocytes and isolated thymocyte nuclei by agents that simulate protein kinase C. J Biol Chem 264:13399-13402

McConkey DJ, Nicotera P, Orrenius S (1994) Signalling and chromatin fragmentation in thymocyte apoptosis. Immunol Rev 142:343-363

McCoy C, Smith DE, Cornwell MM (1995) 12-O-tetradecanoylphorbol-13-acetate activation of the MDR1 promoter is mediated by EGR1. Mol Cell Biol 15:6100-6108

McCoy C, McGee SB, Cornwell MM (1999) The Wilms' tumor suppressor, WT1, inhibits 12-O-tetradecanoylphorbol-13-acetate activation of the multidrug resistance-1 promoter. Cell Growth Differ 10:377-386

McDonald III ER, Wu GS, Waldman T, El-Deiry WS (1996) Repair Defect in p21 WAF1/CIP1 -/- human cancer cells. Cancer Res 56:2250-2255

McDonald AC, Propper D, King D, Champain K, Graf P, Man A, Caponigro F, Thavasu P, Balkwill F, Twelves C, Kaye SB (1997) Phase I and pharmacokinetic study of CGP 41251, an inhibitor of protein kinase C. Proc Am Soc Clin Oncol 16:742

McGahon A, Bissonnette R, Schmitt M, Cotter KM, Green DR, Cotter TG (1994) BCR-ABL maintains resistance of chronic myelogenous leukemia cells to apoptotic cell death. Blood 83:1179-1187

Meggio F, Donella Deana A, Ruzzene M, Brunati AM, Cesaro L, Guerra B, Meyer T, Mett H, Fabbro D, Furet P, Dobrowolska A, Pinna LA (1995) Different susceptibility of protein kinases to staurosporine inhibition. Kinetic studies and molecular bases for resistance of protein kinase CK2. Eur J Biochem 234:317-322

Megidish T, Mazurek N (1989) A mutant protein kinase C that can transform fibroblasts. Nature 342:807-811

Mehta AB, Virchis AE, Jones DT, Hart SM, Wickremasinghe RG, Prentice HG, Young KL, Man A, Csermak-Renner K, Ganeshaguru K (1997) Preliminary results of a phase 2 study on protein kinase C inhibitor (CGP 41251) in low grade lymphoproliferative disorders. Blood 90 (Suppl. 1): 3646

Mellor H, Parker PJ (1998) The extended protein kinase C superfamily. Biochem J 332:281-292

Merrit JE, Sullivan JA, Drew L, Khan A, Wilson K, Mulqueen M, Harris W, Bradshaw D, Hill CH, Rumsby M, Warr R (1999) The bisindolylmaleimide protein kinase C inhibitor, Ro 32-2241, reverses multidrug resistance in KB tumour cells. Cancer Chemother Pharmacol 43:371-378

Meyer T, Regenass U, Fabbro D, Alteri E, Rösel J, Müller M, Carvatti G, Matter A (1989) A derivative of staurosporine (CGP 41251) shows selectivity for PKC inhibition and in vitro antiproliferative as well as in vivo antitumor activity. Int J Cancer,43:851-856

Migliorati G, Nicoletti I, D'Adamio F, Spreca A, Pagliacci C, Riccardi C (1994) Dexamethasone induces apoptosis in mouse natural killer cells and cytotoxic T lymphocytes. Immunology 81:21-26

Militante JD, Lombardini JB (1999) Taurine uptake activity in the rat retina: protein kinase C-independent inhibition by chelerythrine. Brain Res 818:368-374

Miller DS, Sussman CR, Refro JL (1998) Protein kinase C regulation of P-glycoprotein-mediated xenobiotic secretion in renal proximal tubule. Am J Physiol 275:F785-795

Mischak H, Goodnight JA, Kölch W, Martiny-Baron G, Schaechtele C, Kazanietz MG, Blumberg PM, Pierce JH, Mushinski JF (1993) Overexpression of protein kinase C-delta and -epsilon in NIH 3T3 cells induces opposite effects on growth, morphology, anchorage dependence, and tumorigenicity. J Biol Chem 268:6090-6096

Mishima K, Ohno S, Shitara N, Yamaoka K, Suzuki K (1994) Opposite effects of the overexpression of protein kinase C gamma and delta on the growth properties of human glioma cell line U251 MG. Biochem Biophys Res Commun 201:363-372

Misra-Press A, Fields AP, Samols D, Goldthwait DA (1992) Protein kinase C isoforms in human glioblastoma cells. Glia 6:188-197

Miyamoto K, Wakusawa S, Inoko.K, Takagi K, Koyama M (1992) Reversal of vinblastine resistance by a new staurosporine derivative, NA-382, in P388/ADR cells. Cancer-Lett 64:177-183

Miyamoto K, Inoko K, Wakusawa S, Kajita S, Hasegawa T, Takagi K, Masao K (1993) Inhibition of multidrug resistance by a new staurosporine derivative, NA-382, in vitro and in vivo. Cancer Res 53:1555-1559

Mizuno K, Noda K, Ueda Y, Hanaki H, Saido TC, Ikuta T, Kuroki T, Tamaoki T, Hirai S, Osada S, Ohno S (1995) UCN-01, an antitumor drug, is a selective inhibitor of the conventional PKC subfamily FEBS Lett 359:259-261

Mochley-Rosen D, Gordon AS (1998) Anchoring proteins for protein kinase C: a means for isoenzyme selectivity. FASEB J 12:35-42

Montaner S, Ramos A, Perona R, Esteve P, Carnero A, Lacal JC (1995) Overexpression of PKC zeta in NIH3T3 cells does not induce cell transformation nor tumorigenicity and does not alter NF kappa B activity. Oncogene 10:2213-2220

Moore NC, Jenkinson EJ, Owen JJ (1992) Effects of the thymic microenvironment on the response of thymocytes to stimulation. Eur J Immunol 22:2533-2537

Moreland NJ, Illand M, Kim YT, Paul J, Brown R (1999) Modulation of drug resistance mediated by loss of mismatch repair by the DNA polymerase inhibitor aphidicolin. Cancer Res 59:2102-2106

Morgan PF, Fine RL, Montgomery P, Marangos PJ (1991) Multidrug resistance in MCF-7 human breast cancer cells is associated with increased expression of nucleoside transporters and altered uptake of adenosine. Cancer Chemother Pharmacol 29:127-132

Morris CM, Smith GJ (1992) Altered levels and protein kinase C-mediated phosphorylation of substrates in normal and transformed mouse lung epithelial cells. Exp Cell Res 200:149-155

Morse-Gaudio M, Connolly JM, Rose DP (1998) Protein kinase C and its isoforms in human breast cancer cells: relationship to the invasive phenotype. Int J Oncol 12:1349-1354

Moscow JA (1998) Methotrexate transport and resistance. Leuk Lymphoma 30:215-224

Motyka B, Griebel PJ, Reynolds JD (1993) Agents that activate protein kinase C rescue sheep ileal Peyer's patch B cells from apoptosis. Eur J Immunol 23:1314-1321

Muller G, Ayoub M, Storz P, Rennecke J, Fabbro D, Pfizenmaier K (1995) PKC zeta is a molecular switch in signal transduction of TNF-alpha, bifunctionally regulated by ceramide and arachidonic acid. EMBO J 14:1961-1969

Murray NR, Baumgardner GP, Burns DJ, Fields AP (1993) Protein kinase C isotypes in human erythroleukemia (K562) cell proliferation and differentiation. Evidence that beta II protein kinase C is required for proliferation. J Biol Chem 268:15847-15853

Murray NR, Fields AP (1997) Atypical protein kinase C ι protects human leukemia cells against drug-induced apoptosis. J Biol Chem 272:27521-27524

Murray NR, Thompson LJ, Fields AP (1997) The role of protein kinase C in cellular proliferation and cell cycle control. In: Molecular Biology Intelligence Unit: Protein Kinase C, PJ Parker and LV Dekker (eds.), R.G. Landes Company, Austin, TX, Springer-Verlag, Heidelberg, 97-120

Nakamura S, Wakusawa S, Tajima K, Miyamoto K, Hagiwara M, Hidaka H (1993) Effects of isoquinolinesulphonamide compounds on multidrug-resistant P388 cells. J Pharm Pharmacol 45:268-273

Nemunaitis J, Holmlund JT, Kraynak M, Richards D, Bruce J, Ognoskie N, Kwoh TJ, Geary R, Dorr A, Von Hoff D, Eckhardt SG (1999) Phase I evaluation of ISIS 3521, an antisense oligodeoxynucleotide ot protein kinase C-alpha, in patients with advanced cancer. J Clin Oncol 17: 3586-3595

Newton AC, Johnson JE (1998) Protein kinase C: a paradigm for regulation of protein function by two membrane-targeting modules. Biochim Biophys Acta 1376:155-172

Ng T, Squire A, Hansra G, Bornancin F, Prevostel C, Hanby A, Harris W, Barnes D, Schmidt S, Mellor H, Bastiaens PI, Parker PJ (1999) Imaging protein kinase Calpha activation in cells. Science 283:2085-2089

Niedel JE, Kuhn L., Vandenbark GR (1983) Phorbol diester receptor copurifies with protein kinase C. Proc Natl Acad Sci USA 80:36-40

Niimi S, Nakagawa K, Yokota J, Tsunokawa Y, Nishio K, Tershima Y, Shibuya M, Terada M, Saijo N (1991) Resistance to anticancer drugs in NIH3T3 cells transfected with c-myc and/or c-H-ras genes. Int J Cancer 63:237-241

Nishizuka Y (1995) Protein kinase C and lipid signaling for sustained cellular responses. FASEB J 9:484-496

Nixon JS (1997) The biology of protein kinase C inhibitors. In: Molecular Biology Intelligence Unit: Protein Kinase C, PJ Parker and LV Dekker (eds.), R.G. Landes Company, Austin, TX, Springer-Verlag, Heidelberg, 205-236

Noe V, Ciudad CJ (1995) Protein kinase C inhibitors reduce phorbol ester-induced resistance to methotrexate in Chines hamster ovary cells. Biochem Pharmacol 50:337-346

Noseda A, Berens ME, White JG, Modest EJ (1988) In vitro antiproliferative activity of combinations of ether lipid analogues and DNA interactive agents against human tumor cells. Cancer Res 48:1788-1791

Obeid LM, Linardic CM, Karolak LA, Hannun YA (1993) Programmed cell death induced by ceramide. Science 259:1769-1771

O'Brian CA, Vogel, VG, Singletary SE, Ward NE (1989a) Elevated protein kinase C expression in human breast tumor biopsies relative to normal breast tissue. Cancer Res 49:3215-3217

O'Brian CA, Fan D, Ward NE, Seid C, Fidler IJ (1989b) Level of protein kinase C activity correlates directly with resistance to adriamycin in murine fibrosarcoma cells. FEBS Lett 246:78-82

O'Brian CA, Fan D, Ward NE, Dong Z, Iwamoto L, Gupta KP, Earnest LE, Fidler IJ (1991a) Transient enhancement of multidrug resistance by the bile acid deoxycholate in murine fibrosarcoma cells in vitro.Biochem Pharmacol Mar 1;41(5):797-806

O'Brian CA, Ward NE, Liskamp RM, de-Bont DB, Earnest LE, van Boom JH, Fan D (1991b) A novel N-myristylated synthetic octapeptide inhibits protein kinase C activity and partially reverses murine fibrosarcoma cell resistance to adriamycin. Invest New Drugs 9:169-179

O'Brian CA, Ward NE, Gravitt KR, Gupta KP (1995) The tumor promoter receptor protein kinase C: a novel target for chemoprevention and therapy of human colon cancer. Prog Clin Biol Res 391:117-120

Ochs K, Sobol RW, Wilson SH, Kaina B (1999) Cells deficient in DNA polymerase beta are hypersensitive to alkylating agent-induced apoptosis and chromosomal breakage. Cancer Res 59:1544-1551

O'Driscoll KR, Madden PV, Christiansen KM, Viage A, Slaga TJ, Fabbro D, Powell CT, Weinstein IB (1994) Overexpression of protein kinase C beta I in a murine keratinocyte cell line produces effects on cellular growth, morphology and differentiation. Cancer Lett 83:249-259

O'Driscoll KR, Teng KK, Fabbro D, Greene LA, Weinstein IB (1995) Selective translocation of protein kinase C-delta in PC12 cells during nerve growth factor-induced neuritogenesis. Mol Biol Cell 6:449-458

Ohmi Y, Ohta A, Sasakura Y, Sato N, Yahata T, Santa K, Habu S, Nishimura T (1997) The role of phorbol ester-sensitive protein kinase C isoforms in lymphokine-activated killer cell-mediated cytotoxicity: dissociation between perforin-dependent and Fas-dependent cytotoxicity. Biochem Biophys Res Commun 235:461-464

Ohmichi M, Zhu G, Saltiel AR (1993) Nerve growth factor activates calcium-insensitive protein kinase C-epsilon in PC-12 rat pheochromocytoma cells. Biochem J 295 (Pt 3):767-772

Ohmori T, Arteaga CL (1998) Protein kinase C epsilon translocation and phosphorylation by cis-diamminedichloroplatinum(II) (CDDP): potential role in CDDP-mediated cytotoxicity. Cell Growth Differ 9:345-353

Ojeda F, Guarda MI, Maldonado C, Folch H, Diehl H (1992) Role of protein kinase-C in thymocyte apoptosis induced by irradiation. Int J Radiat Biol 61:663-667

Oka M, Ogita K, Ando H, Horikawa T, Hayashibe K, Saito N, Kikkawa U, Ichihashi M (1996) Deletion of specific protein kinase C subspecies in human melanoma cells. J Cell Physiol 167:406-412

Omura S, Iwai Y, Hirano A, Nakagawa A, Awaya J, Tsuchya H, Takahashi Y, Masuma R (1977) A new alkaloid AM-2282 OF Streptomyces origin. Taxonomy, fermentation, isolation and preliminary characterization. J Antibiot 30:275-282

Ozols RF, O'Dwyer PJ, Hamilton TC, Young RC (1990) The role of glutathione in drug resistance. Cancer Treat Rev 17A:45-50

Perego P, Casati G, Gambetta RA, Soranzo C, Zunino F (1993) Effect of modulation of protein kinase C activity on cisplatin toxicity in cisplatin-resistant and cisplatin-sensitive human osteosarcoma cells. Cancer Lett, 72:53-58

Perez RP, Hamilton TC, Ozols RF (1990) Resistance to alkylating agents and cisplatin: insights from ovarian carcinoma model systems. Pharmac Ther 48:19-27

Perletti GP, Folini M, Lin HC, Mischak H, Piccinini F, Tashjian AH Jr (1996) Overexpression of protein kinase C epsilon is oncogenic in rat colonic epithelial cells. Oncogene 12:847-854

Perletti GP, Marras E, Concari P, Piccinini F, Tshjian AH Jr (1999) PKC delta acts as a growth and tumor suppressor in rat colonic epithelial cells. Oncogene 18:1251-1256

Perotti M, Toddei F, Mirabelli F, Vairetti M, Bellomo G, McConkey DJ, Orrenius S (1990) Calcium-dependent DNA fragmentation in human synovial cells exposed to cold shock. FEBS Lett 259:331-334

Persons DA, Wilkinson,WO, Bell RM, Finn OJ (1988) Altered growth regulation and enhanced tumorigenicity of NIH 3T3 fibroblasts transfected with protein kinase C-I cDNA. Cell 52:447-458

Pettit GR, Day JF, Hartwell JL, Wood HB (1970) Antineoplastic components of marine animals. Nature 227:962-963

Phan SC, Morotomi M, Guillem JG, LoGerfo P, Weinstein IB (1991) Decreased levels of 1,2-sn-diacylglycerol in human colon tumors. Cancer Res 51:1571-1573

Piovesan B, Pennell N, Berinstein NL (1998) Human lymphoblastoid cell lines expressing mutant p53 exhibit decreased sensitivity to cisplatin-induced cytotoxicity. Oncogene 17:2339-2350

Pizao PE, Peters GJ, Van Ark-Otte J, Smets LA, Smitskamp-Wilms E, Winograd B, Pinedo HM, Giaccone G (1993) Cytotoxic effects of anticancer agents on subconfluent and multilayered postconfluent cultures. Eur J Cancer 29A:1566-1573

Pollak IF, Kawecki S, Lazo JS (1996) Blocking of glioma proliferation in vitro and in vivo and potentiating the effects of BCNU and cis-platin: UCN-01, a selective protein kinase C inhibitor. J Neurosurg 84:1024-1032

Pongracz J, Clark P, Neoptolemous JP, Lord JM (1995a) Expression of protein kinase C isoenzymes in colorectal cancer tissue and their differential activation by different bile acids. Int J Cancer 61:35-39

Pongracz J, Tuffley W, Johnson GD, Deacon EM, Burnett D, Stockley RA, Lord JM (1995b) Changes in protein kinase C isoenzyme expression associated with apoptosis in U937 myelomonocytic cells. Exp Cell Res 218:430-438

Pongracz J, Deacon EM, Johnson GD, Burnett D, Lord JM (1996) Doppa induces cell death but not differentiation of U937 cells: evidence for the involvement of PKC-beta 1 in the regulation of apoptosis. Leuk Res 20:319-326

Posada J, Vichi P, Tritton TR (1989a) Protein kinase C in adriamycin action and resistance in mouse sarcoma 180 cells. Cancer Res 49:6634-6639

Posada JA, McKeegan EM, Worthington KF, Morin MJ, Jaken S, Tritton TR (1989b) Human multidrug resistant KB cells overexpress protein kinase C: involvement in drug resistance. Cancer Commun 1:285-292

Powell CT, Brittis NJ, Stec D, Hug H, Heston WD, Fair WR (1996) Persistent membrane translocation of protein kinase C alpha during 12 0 tetradecanoylphorbol 13 acetate induced apoptosis of LNCaP human prostate cancer cells. Cell Growth Differ 7:419 428

Powis G (1991) Signalling targets for anticancer drug development. Trends Pharmacol Sci 12:188-194

Prendiville J, Crowther D, Thatcher N, Woll PJ, Fox BW, McGown A, Testa N, Stern P, McDermott R, Potter M, Pettit GR (1993) A phase I study of intravenous bryostatin 1 in patients with advanced cancer. Br J Cancer 68:418-424

Prendiville J, McGown AT, Gescher A, Dickson AJ, Courage C, Pettit GR, Crowther D, Fox BW (1994) Establishment of a murine leukaemia cell line resistant to the growth-inhibitory effect of bryostatin 1. Br J Cancer 70: 573-578

Prevostel C, Alvaro V, de-Boisvilliers F, Martin A, Jaffiol C, Joubert D (1995) The natural protein kinase C alpha mutant is present in human thyroid neoplasms. Oncogene 11:669-674

Prevostel C, Martin A, Alvaro V, Jaffiol C, Joubert D (1997) Protein kinase C alpha and tumorigenesis of the endocrine gland. Horm Res 47:140-144

Propper DJ, Macaulay V, O'Byrne KJ, Braybrooke JP, Wilner SM, Ganesan TS, Talbot DC, Harris AL (1998) A phase II study of bryostatin 1 in metastatic malignant melanoma. Br J Cancer 78:1337-1341

Rahmsdorf HJ, Herrlich P (1990) Regulation of gene expression by tumor promoters. Pharmac Ther 48:157-188

Rajotte D, Haddad P, Haman A, Cragoe EJJr, Hoang T (1992) Role of protein kinase C and the Na$^+$/H$^+$ antiporter in suppression of apoptosis by granulocyte macrophage colony-stimulating factor and interleukin-3. J Biol Chem 267:9980-9987

Rea D, Prendiville J, Harris AL, Woll P, Philip P, Carmichael J, Lewis C, Dallma M, Walker T, Sri-Pathmanathan E, Thatcher N, Crowther D (1992) A phase I study of bryostatin 1 – a PKC partial agonist. Ann Ancol 3:62

Reddig PJ, Dreckschmidt NE, Ahrens H, Simsiman R, Tseng C-P, Zou J, Oberley TD, Verma AK (1999) Transgenic mice overexpressing protein kinase C δ in the epidermis are resistant to skin tumor promotion by 12-O-tetradecanoylphorbol-13-acetat, Cancer Res 49:5710-5718

Regev R, Assaraf YG, Eytan GD (1999) Membrane fluidization by ehter, other anesthetics, and certain agents abolishes P-glycoprotein ATPase activity and modulates efflux from multidrug-resistant cells. Eur J Biochem 259:18-24

Rennecke J, Rehberger PA, Fürstenberger G., Johannes FJ, Stöhr M, Marks F, Richter KH (1999) Protein-kinase-Cμ expression correlates with enhanced kerationcate proliferation in normal and neoplastic mouse epidermis and in cell culture. Int J Cancer 80:98-103

Riley RJ, Workman P (1992) DT-diaphorase and cancer chemotherapy. Biochem Pharmacol 43:1657-1669

Robert J, Larsen AK (1998) Drug resistance to topoisomerase II inhibitors. Biochimie 80:247-254

Robinson LJ, Roberts WK, Ling TT, Lamming D, Sternberg SS, Roepe PD (1997) Human MDR 1 protein overexpression delays the apoptotic cascade in Chinese hamster ovary fibroblasts. Biochemistry 36:11169-11178.

Rodriguez-Pena A, Rozengurt E (1984) Disappearance of Ca^{2+}-sensitive, phospholipid-dependent protein kinase activity in phorbol ester-treatet 3T3 cells. Biochem Biophys Res Comm 120:1053-1059

Romanova LY, Alexandrov IA, Schwab G, Hilbert DM, Mushinski JF, Nordan RP (1996) Mechanism of apoptosis suppression by phorbol ester in IL-6-starved murine plasmacytomas: role of PKC modulation and cell cycle. Biochemistry 35:9900-9906

Rosales OR, Isales CM, Bhargava J (1998) Overexpression of protein kinase C alpha and beta1 has distinct effects on bovine aortic endothelial cell growth. Cell Signal 10:589-597

Ross DD, Yang W, Abruzzo LV, Dalton, WS, Schneider E, Lage H, Dietel M, Greenberger L, Cole SPC, Doyle AL (1999) Atypical multidrug resistance: breast cancer resistance protein messenger RNA expression in mitoxanthrone-selected cell lines. J Natl Cancert Inst 91:429-433

Ruiz-Ruiz MC, Izquierdo M, de Murcia G, Lopez-Rivas A (1997) Activation of protein kinase C attenuates early signals in Fas-mediated apoptosis. Eur J Immunol 27:1442-1450

Rusnak JM, Lazo JS (1996) Downregulation of protein kinase C suppresses induction of apoptosis in humanprostatic carcinoma cells. Exp Cell Res 224:189-99

Ruvolo PP, Deng X, Carr BK, May WS (1998) A functional role for mitochondrial protein kinase Calpha in Bcl2 phosphorylation and suppression of apoptosis. J Biol Chem 273:25436-25442

Sachs CW, Safa AR, Harrison SD, Fine RL (1995) Partial inhibition of multidrug resistance by safingol is independent of modulation of P-glycoprotein substrate activities and correlated with inhibition of protein kinase C. J Biol Chem 270: 26639-26648

Sachs CW, Ballas LM, Mascarella SW, Safa AR, Lewin AH, Loomis C, Carroll FI, Bell RM, Fine-RL (1996) Effects of sphingosine stereoisomers on P-glycoprotein phosphorylation and vinblastine accumulation in multidrug-resistant MCF-7 cells. Biochem Pharmacol 52:603-612

Sakakura C, Sweeney EA, Shirahama T, Hakomori S, Igarashi Y (1996) Suppression of bcl-2 gene expression by sphingosine in the apoptosis of human leukemic HL-60 cells during phorbol ester-induced terminal differentiation. FEBS Lett 379:177-180

Sakata K, Kwok TT, Murphy BJ, Laderoute KR, Gordon GR, Sutherland RM (1991) Hypoxia-induced drug resistance: comparison to P-glycoprotein-associated drug resistance. Br J Cancer 64:809-814

Sakurada K, Zheng B, Kuo JF (1992) Comparative effects of protein phosphatase inhibitors (okadaic acid and calyculin A) on human leukemia HL60, HL60/ADR and K562 cells. Biochem Biophys Res Commun 187:488-492

Sampson E, Wolff CL, Abraham I (1993) Staurosporine reduces P-glycoprotein expression and modulates multidrug resistance. Cancer Lett 68:7-14

Sanchez V, Lucas M, Sanz A, Goberna R (1992) Decreased protein kinase C activity is associated with programmed cell death (apoptosis) in freshly isolated rat hepatocytes. Biosci Rep 12:199-206

Sanchez-Margalet V, Lucas M, Solano F, Goberna R (1993) Sensitivity of insulin-secreting RIN m5F cells to undergoing apoptosis by the protein kinase C inhibitor staurosporine. Exp Cell Res 209:160-163

Sanchez-Perez I, Perona R (1999) Lack of c-Jun activity increases survival to cisplatin. FEBS Lett 453:151-158

Sanchez-Prieto R, Vargas JA, Carnero A, Marchetti E, Romero J, Durantez A, Lacal JC, Ramon-y-Cajal S (1995) Modulation of cellular chemoresistance in keratinocytes by activation of different oncogenes. Int J Cancer 60:235-243

Sato W, Yusa K, Naito M, Tsuruo T (1990) Staurosporine, a potent inhibitor of C-kinase, enhances drug accumulation in multidrug-resistant cells. Biochem Biophys Res Comm 173:1252-1257

Sausville EA, Lush RD, Headlee D, Smith AC, Figg WD, Arbuck SG, Senderowicz AM, Fuse E, Tanii H, Kuwabara T, Kobayashi S (1998) Clinical pharmacology of UCN-01: initial observations and comparison to preclinical models. Cancer Chemother Pharmacol 42, Suppl:S54-59

Saxon ML, Zhao X, Black JD (1994) Activation of protein kinase C isozymes is associated with post-mitotic events in intestinal epithelial cells in situ. J Cell Biol 126:747-763

Scaglione-Sewell B, Abraham C, Bissonnette M, Skarosi SF, Hart J, Davidson NO, Wali RK, Davis BH, Sitrin M, Brasitus TA (1998) Decreased PKC-alpha expression increases cellular proliferation, decreases differentiation, and enhances the transformed phenotype of CaCo-2 cells. Cancer Res 58:1074-1081

Scala S, Dickstein B, Regis J, Szallasi Z, Blumberg PM, Bates SE (1995) Bryostatin 1 affects P-glycoprotein phosphorylation but not function in multidrug-resistant human breast cancer cells. Clinical Cancer Res 1:15851-1587

Scanlon, KJ, Kashani-Sabet M, Tone T, Funato T (1991) Cisplatin resistance in human acancers. Pharmac Ther 52:385-406

Scheffer GL, Wijngaard PL, Flens MJ, Izquierdo MA, Slovak ML, Pinedo HM, Meijer CJ, Clevers HC, Scheper RJ (1995) The drug resistance-related protein LRP is the human major vault protein. Nat Med 1:578-582

Schiemann U, Assert R, Moskopp D, Gellner R, Hengst K, Gullotta F, Domschke W, Pfeiffer A (1997) Analysis of a protein kinase C-alpha mutation in human pituitary tumours. J Endocrinol 153:131-137

Schimke RT (1986) Methotrexate resistance and gene amplification. Mechanisms and implications. Cancer 57:1912-1917

Schlatterer K, Krauter G, Schlatterer B, Hecker E, Chandra P (1999) A novel protein (p10) induced by 12-O-tetradecanoyl-phorbol-13-acetate (TPA) and other hyperplasiogenic tumor-promoting and non-promoting agents in murine epidermis. Anticancer Res 19:397-404

Schönwasser DC, Marais RM, Marshall CJ, Parker PJ (1998) Activation of the mitogen-activated protein kinase/extracellular signal-regulated kinase pathway by conventional, novel, and atypical protein kinase C isotypes. Mol Cell Biol 18: 90-798

Schwartz GK, Arkin H, Holland JF, Ohnuma T (1991) Protein kinase C activity and multidrug resistance in MOLT-3 human lymphoblastic leukemia cells resistant to trimetrexate. Cancer Res 51:55-61

Schwartz GK, Haimovitz Friedman A, Dhupar SK, Ehleiter D, Maslak P, Lai L, Loganzo F Jr, Kelsen DP, Fuks Z, Albino AP (1995) Potentiation of apoptosis by treatment with the protein kinase C specific inhibitor safingol in mitomycin C treated gastric cancer cells. J Natl Cancer Inst 87:1394-1399

Segal-Bendirdjian E, Jacquemin-Sablon A (1996) Cisplatin resistance in a murine leukemia cell line associated with defect of apoptosis. Bull Cancer 83:371-378

Sewing A, Wiseman B, Lloyd AC, Land H (1997) High-intensity raf signal causes cell cycle arrest mediated by p21Cip1. Mol Cell Biol 17: 5588-5597

Seynaeve CM, Stetler-Stevenson M, Sebers S, Kaur G, Sausville EA, Worland PJ (1993) Cell cycle arrest and growth inhibition by the protein kinase antagonist UCN-01 in human breast carcinoma cells. Cancer Res 53:2081-2086

Seynaeve CM, Kazanietz MG, Blumberg PM, Sausville AE, Worland PJ (1994) Differential inhibition of protein kinase C isozymes by UCN-01, a staurosporine analogue. Mol Pharmacol 45:1207-1214

Sha EC, Sha MC, Kaufmann SH (1996) Evaluation of 2,6-diamino-N-([1-(1-oxotridecyl)-2-piperidinyl]methyl)-hexanamide (NPC 15437), a protein kinase C inhibitor, as a modulator of P-glycoprotein-mediated resistance in vitro. Invest New Drugs 13:285-294

Shao RG, Cao CX, Pommier Y (1997) Activation of PKC alpha downstream from caspases during apoptosis induced by 7-hydroxystaurosporine or the topoisomerase inhibitors, camptothecin and etoposide, in human myeloid leukemia HL60 cells. J Biol Chem 272:31321-31325

Shaposhnikova VV, Dobrovinskaya OR, Eidus LKh, Korystov YN (1994) Dependence of thymocyte apoptosis on protein kinase C and phospholipase A2. FEBS Lett 348:317-319

Shen L, Glazer RI (1998) Induction of apoptosis in glioblastoma cells by inhibition of protein kinase C and its association with the rapid accumulation of p53 and induction of the insulin-like growth factor-1-binding-protein-3. Biochem Pharmacol 55:1711-1719

Shimizu E, Zhao MR, Nakanishi H, Yamamoto A, Yoshida S, Takada M, Ogura T, Sone S (1996) Differing effects of staurosporine and UCN-01 on RB protein phosphorylation and expression of lung cancer cell lines. Oncology 53:494-504

Shinoura N, Yoshida Y, Asai A, Kirino T, Hamada H (1999) Relative level of expression of bax and bcl-XL determines the cellular fate of apoptosis/necrosis induced by the overexpression of Bax. Oncogene 18:5703-5713

Shirahama T, Sakakura C, Sweeney EA, Ozawa M, Takemoto M, Nishiyama K, Ohi Y, Igarashi Y (1997) Sphingosine induces apoptosis in androgen-independent human prostatic carcinoma DU-145 cells by suppression of bcl-X(L) gene expression. FEBS Lett 407:97-100

Shoji M, Raynor RL, Berdel WE, Vogler WR, Kuo JF (1988) Effects of thioetherphospholipid BM41440 on PKC and phorbol ester-induced differentiation of human leukemia HL60 and KG-1 cells. Cancer Res 48:6669-6673

Shoji M, Raynor RL, Fleer EA, Eibl H, Vogler WR, Kuo JF (1991) Effects of hexadecylphosphocholine on protein kinase C and TPA-induced differentiation of HL60 cells. Lipids 26:145-149

Siemann DW, Jiang JB, Ballas L, Janzen W (1993) Threo-dihydrosphingosine potentiates the in vivo antitumor efficacy of cisplatin and adriamycin. Proc Am Assoc Cancer Res 34: 410, 2452

Sklar MD, Prochownik EV (1991) Modulation of cisplatinum resistance in friend erythroleukemia cells by c-myc. Cancer Res 51:2118-2123

Slapak CA, Daniel JC, Levy SB (1990) Sequential emergence of distinct resistance phenotypes in murine erythroleukemia cells under adriamycin selection: decreased anthracycline uptake precedes increase P-glycoprotein expression. Cancer Res 50:7895-7901

Smith CD, Zilfou JT (1995) Circumvention of P-glycoprotein-mediated multiple drug resistance by phosphorylation modulators is independent of protein kinases. J Biol Chem 270:28145-28152

Smith JB, Smith L, Pettit GR (1985) Bryostatins: potent new mitogens that mimic phorbol ester tumor promoters. Biochem. Biophys Res Commun 132:939-945

Smyth MJ, Drasovskis E, Sutton VR, Johnstone RW (1998) The drug efflux protein, P-glycoprotein, additionally protects drug-resistant tumor cells from multiple forms of caspase-dependent apoptosis. Proc Natl Acad Sci USA 95:7024-7029

Solary E, Bertrand R, Kohn KW, Pommier Y (1993) Differential induction of apoptosis in undifferentiated and differentiated HL-60 cells by DNA topoisomerase I and II inhibitors. Blood 81:1359-1368

Sommers GM, Alfieri AA (1998) Multimodality therapy: radiation and continuous concomitant cis-platinum and PKC inhibition in a cervical carcinoma model. Cancer Invest 16:462-470

Spitaler M, Utz I, Hilbe W, Hofmann J, Grunicke HH (1998) PKC-independent modulation of multidrug resistance in cells with mutant (V185) but not wild-type (G185) P-glycoprotein by bryostatin 1. Biochem Pharmacol 56:861-869

Spitaler M, Wiesenhofer B, Biedermann V, Seppi T, Zimmermann J, Grunicke H, Hofmann J (1999) The involvement of protein kinase C isoenzymes α, ε and ζ on the sensitivity to antitumor treatment and apoptosis induction. Anticancer Res 19:3969-3976

Staats J, Marquardt D, Center MS (1990) Characterization of a membrane-associated protein kinase of multidrug-resistant HL60 cells which phosphorylates P-glycoprotein. J Biol Chem 265:4084-4090

Stabel S, Parker PJ (1991) Protein kinase C. Pharmac Ther 41:71-95

Stabel S (1994) Protein kinase C - an enzyme and its relatives. Sem Cancer Biol 5:277-284

Staroselsky AN, Fan D, O'Brian CA, Bucana CD, Gupta KP, Fidler IJ (1990) Site-dependent differences in response of the UV-2237 murine fibrosarcoma to systemic therapy with adriamycin. Cancer Res 50:7775-7780

StCroix B, Florenes VA, Rak JW, Flanagan M, Bhattacharya N, Slingerland JM, Kerbel RS (1996) Impact of the cyclin-dependent kinase inhibitor p27Kip1 on resistance of tumor cells to anticancer agents. Nat Med 2:1204-1210

StCroix B, Kerbel RS (1997) Cell adhesion and drug resistance in cancer. Curr Opin Oncol 9:549-556

Stevens VL, Nimkar S, Jamison WC, Liotta DC, Merrill AJJr (1990) Characteristics of the growth inhibition and cytotoxicity of long-chain (sphingoid) bases for Chinese hamster ovary cells: evidence for an involvement of protein kinase C. Biochim Biophys Acta 1051:37-45

Stone RM, Sariban E, Pettit GR, Kufe DW (1988) Bryostatin 1 activates PKC and induces monocytic differentiation of HL60 cells. Blood 72:208-213.

Stromskaya TP, Grigorian IA, Ossovskaya VS, Rybalkina EY, Chumakov PM, Kopnin BP (1995) Cell-specific effects of RAS oncogene and protein kinase C agonist TPA on P-glycoprotein function. FEBS Lett 368:373-376

Suga K, Sugimoto I, Ito H, Hashimoto E (1998) Down-regulation of protein kinase C-alpha detected in human colorectal cancer. Biochem Mol Biol Int 44:523-528

Sun XG, Rotenberg SA (1999) Overexpression of protein kinase Calpha in MCF-10A human breast cells engenders dramatic alterations in morphology, proliferation, and motility. Cell Growth Differ 10:343-352

Susick, RL, Bozigian HP, Kreutzberg J, Adams LM, Weiler MS, Harrison SD, Kedderis LB (1993) Combination toxicology studies with the chemopotentiating agent SPC-100270 (a PKC inhibitor) and chemotherapeutic agents. Proc Am Assoc Cancer Res 34:410, 2444

Sutherland RM (1988) Cell and environment interactions in tumor microregions: the multicell speroid model. Science 240:177-184

Szallasi Z, Smith CB, Pettit GR, Blumberg PM (1994) Differential regulation of protein kinase C isozymes by bryostatin 1 and phorbol 12-myristate 13-acetate in NIH 3T3 fibroblasts. J Biol Chem 269:2118-2124

Szallasi Z, Du L, Levine R, Lewin NE, Nguyen PN, Williams MD, Pettit GR, Blumberg PM (1996) The bryostatins inhibit growth of B16/F10 melanoma cells in vitro through a protein kinase C-independent mechanism: dissociation of activities using 26-epi-bryostatin 1. Cancer Res 56(9):2105-11

Tamaoki T, Nomoto H, Takahashi I, Kato Y, Morimoto M, Tomita F (1989) Staurosporine, a potent inhibitor of phospholipid/Ca^{2+}-dependent protein kinase. Biochem Biophys Res. Commun 135:397-402

Teofili L, Pierelli L, Iovino MS, Leone G, Scambia G, De Vincenzo R, Benedetti-Panici P, Menichella G, Macri E, Piantelli M, Ranelletti FO, Larocca LM (1992) The combination of quercetin and cytosine arabinoside synergistically inhibits leukemic cell growth. Leuk Res 16:497-503

Tew KD (1994) Glutathione-associated enzymes in anticancer drug resistance. Cancer Res 54:4313-4320

Toullec D, Pianetti P, Coste H, Bellevergue P, Grand-Perret T, Ajakane M, Baudet V, Boissing P, Boursier E, Loriolle F, Duhamel L., Charon F, Kirilovski J (1991) The bis-indolylmaleimide GF109203x is a potent and selective inhibitor of PKC. J Biol Chem 266:15771-15781

Tsai CM, Chang KT, Wu LH, Chen JY, Gazdar AF, Mitsudomi T, Chen MH, Perng RP (1996) Correlations between intrinsic chemoresistance and HER-2/neu gene expression, p53 gene mutations, and cell proliferation characteristics in non-small cell lung cancer cell lines. Cancer Res 56:206-209

Überall F, Oberhuber H, Maly K, Zaknun J, Demuth L, Grunicke HH (1991) Hexadecylphosphocholine inhibits inositol phosphate formation and protein kinase C activity. Cancer Res 51:807-812

Überall F, Hellbert K, Kampfer S, Maly K, Villunger A, Spitaler M, Mwanjewe J, Baier-Bitterlich G, Baier G, Grunicke HH (1999) Evidence that atypical protein kinase C-λ and atypical protein kinase C-ζ participate in ras-mediated reorganization of the F-actin cytoskeleton. J Cell Biol 144:413-425

Uchiumi T, Kohno K, Tanimura H, Hidaka K, Asakuno K, Abe H, Uchida Y, Kuwano M (1993) Involvement of protein kinase in environmental stress-induced activation of human multidrug resistance 1 (MDR1) gene promoter. FEBS Lett 326:11-16

Urasaki Y, Ueda T, Yoshida A, Fukushima T, Takeuchi N, Tsuruo T, Nakamura T (1996) Establishment of a daunorubicin-resistant cell line which shows multidrug resistance by multifactorial mechanisms. Anticancer Res 16:709-714

Utz I, Hofer S, Regenass U, Hilbe W, Thaler J, Grunicke H, Hofmann J (1994) The protein kinase C inhibitor CGP 41251, a staurosporine derivative with antitumor activity, reverses multidrug resistance. Int J Cancer 57:104-110

Utz I, Gekeler V, Ise W, Beck J, Spitaler M, Grunicke H, Hofmann J (1996) Protein kinase C isoenzymes, p53, accumulation of rhodamine 123, glutathione-S-transferase, topoisomerase II and MRP in multidrug resistant cell lines. Anticancer Res 16:289-296

Utz I, Spitaler M, Rybczynska M, Ludescher C, Hilbe W, Regenass U, Grunicke H, Hofmann J (1998) Reversal of multidrug resistance by the staurosporine derivatives CGP 41251 and CGP 42700. Int J Cancer 77:64-69

Valverde AM, Sinnett-Smith J, Van Lint J, Rozengurt E (1994) Molecular cloning and characterization of protein kinase D: a target for diacylglycerol and phorbol esters with a distinctive catalytic domain. Proc Natl Acad Sci USA 91:8572-8576

Varterasian ML, Mohammad RM, Eilender DS, Hulburd K, Rodriguez DH, Pemberton PA, Pluda JM, Dan MD, Pettit GR, Chen BD, Al-Katib AM (1998) Phase I study of bryostatin 1 in patients with relapsed non-Hodgkin's lymphoma and chronic lymphocytic leukemia. Clin Oncol 16:56-62

Vaupel P, Kallinowski F, Okunieff P (1989) Blood flow, oxygen and nutrient supply, and metabolic microenvironment of human tumors: a review. Cancer Res 49:6449-6465

Vogler WR, Whigham E, Bennet WD, Olson AC (1985) Effect of alkyllysophospholipids on phosphatidylcholine biosynthesis in leukemic cell lines. Exp Hematol 13:629-633

Volm M, Pommerenke EW (1995) Associated expression of protein kinase C with resistance to doxorubicin in human lung cancer. Anticancer Res 15:463-466

Wakusawa S, Nakamura S, Tajima K, Miyamoto K, Hagiwara M, Hidaka H (1992) Overcoming of vinblastine resistance by isoquinolinesulfonamide compounds in adriamycin-resistant leukemia cells. Mol Pharmacol 41:1034-1038

Wakusawa S, Inoko K, Miyamoto K (1993) Staurosporine derivatives reverse multidrug resistance without correlation with their protein kinase inhibitory activities. J Antibiotcs 46:335-337

Walter RJ, Shtil AA, Roninson IB, Holian O (1997) 60-Hz electric fields inhibit protein kinase C activity and multidrug resistance gene (MDR1) up-regulation. Radiat Res 147:369-375

Wang Q, Worland PJ, Clark JL, Carlson BA, Sausville EA (1995) Apoptosis in 7-hydroxystaurosporine-treated T lymphoblasts correlates with activation of cyclin-dependent kinases 1 and 2. Cell Growth Differ 6:927-936

Wang J, Walsh K (1996) Resistance to apoptosis conferred by CDK inhibiors during myocyte differentiation. Science 273:359-361

Wang Q, Fan S, Eastman A, Worland PJ, Sausville EA, O'Connor PM (1996) UCN-01: a potent abrogator of G1 checkpoint function in cancer cells with disrupted P53. J Natl Cancer Inst 88:956-965

Wang S, Vrana JA, Bartimole TM, Freemerman AJ, Jarvis WD, Kramer LB, Krystal G, Dent P, Grant S (1997) Agents that down-regulate or inhibit protein kinase C circumvent resistance to 1-beta-D-arabinofuranosylcytosine-induced apoptosis in human leukemia cells that overexpress Bcl-2. Mol Pharmacol 52:1000-1009

Wang XY, Repasky E, Liu HT (1999a) Antisense inhibition of protein kinase C alpha reverses the transformed phenotype in human lung carcinoma cells. Exp Cell Res 250:253-263

Wang CY, Cusack JCJr, Liu R, Baldwin ASJr (1999b) Control of inducible chemoresistance: enhanced anti-tumor therapy through increased apoptosis by inhibition of NF-kappaB. Nat Med 5:412-417

Ward NE, O'Brian CA (1991) Distinct patterns of phorbol ester-induced downregulation of protein kinase C activity in adriamycin-selected multidrug resistant and parental murine fibrosarcoma cells. Cancer Lett 58:189-193

Warenius HM, Seabra LA, Maw P (1996) Sensitivity to cis-diamminedichloroplatinum in human cancer cells is related to expression of cyclin D1 but not c-raf-1 protein. Int J Cancer 67:224-231

Watanabe T, Ono Y, Taniyama Y, Hazama K, Igarashi K, Ogita K, Kikkawa U, Nishizuka Y (1992) Cell division arrest induced by phorbol ester in CHO cells overexpressing protein kinase C-delta subspecies. Proc Natl Acad Sci USA 89:10159-10163

Watts JD, Aebersold R, Polverino AJ, Patterson SD, Gu M (1999) Ceramide second messengers and ceramide assays. Trends Biochem Sci 282:228

Ways DK, Posekany K, deVente J, Garris T, Chen J, Hooker J, Qin W, Cook P, Fletcher D, Parker P (1994) Overexpression of protein kinase C zeta stimulates leukemic cell differentiation. Cell Growth Differ 5:1195-1203

Ways KD, Kukoly CA, deVente J, Hooker JL, Bryant WO, Posekany KJ, Fletcher DJ, Cook PP, Parker PJ (1995) MCF-7 breast cancer cells transfected with protein kinase C-α exhibit altered expression of other protein kinase C isoforms and display a more aggressive phenotype. J Clin Invest 95:1906-1915

Weinstein IB (1991) Nonmutagenic mechanisms in carcinogenesis: Role of protein kinase C in signal transduction and growth control Environmental Health Perspectives 93:175-179

Weisburg JH, Roepe PD, Dzekunov S, Scheinberg DA (1999) Intracellular pH and multidrug resistance regulate complement-mediated cytotoxicity of nucleated human cells. J Biol Chem 274:10877-10888

Westmacott D, Bradshow D, Kumar MKH, Lewis EJ, Murray EJ, Nixon JS, Sedgwick AD (1991) Molecular basis of new approaches to the therapy of rheumatoid arthritis. Molecular Aspects Med 12,395-473

Westphal CH, Hoyes KP, Canman CE, Huang X, Kastan MB, Hendry JH, Leder P (1998) Loss of ATM radiosensitizes multiple p53 null tissues. Cancer Res 58:5637-5639

Whelan RD, Parker PJ (1998) Oncogene Loss of protein kinase C function induces an apoptotic response. Oncogene 16:1939-1944

Whelan RD, Kiley SC, Parker PJ (1999) Tetradecanoyl phorbol acetate-induced microtubule reorganization is required for sustained mitogen-activated protein kinase activation and morphological differentiation of U937 cells. Cell Growth Diff 10:271-277

White E, Prives C (1999) DNA damage enables p73. Natur 399:734-737

Whitman SP, Civoli F, Daniel LW (1997) Protein kinase CbetaII activation by 1-beta-D-arabinofuranosylcytosine is antagonistic to stimulation of apoptosis and Bcl-2alpha down-regulation. J Biol Chem 272:23481-2384

Wielinga PR, Heijn M, Broxterman HJ, Lankelma J (1997) P-glycoprotein-independent decrease in drug accumulation by phorbol ester treatment of tumor cells. Biochem Pharmacol 54:791-799

Williams AC, Collard TJ, Paraskeva C (1999) An acidic environment leads to p53 dependent induction of apoptosis in human adenoma and carcinoma cell lines: implications for clonal selection during colorectal carcinogenesis. Oncogene 18:3199-3204

Wilson RE, Dooley TP, Hart IR (1989) Induction of tumorigenicity and lack of in vitro growth requirement for 12-O-tetradecanoylphorbol-13-acetate by transfection of murine melanocytes with v-Ha-ras. Cancer Res 49:711-716

Woo KR, Shu WP, Kong L, Liu BC (1996) Tumor necrosis factor mediates apoptosis via Ca^{++}/Mg^{++} dependent endonuclease with protein kinase C as a possible mechanism for cytokine resistance in human renal carcinoma cells. J Urol 155:1779-1783

Woodley SL, McMillan M, Shelby J, Lynch DH, Roberts LK, Ensley RD, Barry WH (1991) Myocyte injury and contraction abnormalities produced by cytotoxic T lymphocytes. Circulation 83:1410-1418

Wooten MW, Zhou G, Seibenhener ML, Coleman ES (1994) A role for zeta protein kinase C in nerve growth factor-induced differentiation of PC12 cells. Cell Growth Differ 5:395-403

Workman P (1991) Antitumor ether lipids: endocytosis as a determinant of cellular sensitivity. Cancer Cells 3:315-317

Xiao H, Goldthwait DA, Mapstone T (1994) The identification of four protein kinase C isoforms in human glioblastoma cell lines: PKC alpha, gamma, epsilon, and zeta. J Neurosurg 81:734-740

Xu X, Baltimore D (1996) Dual roles of ATM in the cellular response to radiation and in cell growth. Gene Develop 10:2401-2410

Yang JM, Chin KV, Hait WN (1996) Interaction of P-glycoprotein with protein kinase C in human multidrug resistant carcinoma cells.Cancer Res 56:3490-3494

Yarbro JW (1992) The scientific basis of cancer chemotherapy. In: The Chemotherapy Sourcebook (Perry MC ed.) Williams, Wilkins, 2-8

Yazaki T, Ahmad S, Chahlavi A, Zylber-Katz E, Dean NM, Rabkin SD, Martuza RL, Glazer RI (1996) Treatment of glioblastoma U-87 by systemic administration of an antisense protein kinase C-alpha phosphorothiate oligodeoxynucleotide. Mol Pharmacol 50:236-242

Yen Y, Grill SP, Dutschman GE, Chang C-N, Zhou BS, Cheng Y-C (1994) Characterization of a hydroxyurea-resistant human KB cell line with supersensitivity to 6-thioguanine. Cancer Res 54:3686-3691

Yoshida M, Feng W, Saijo N, Ikekawa T (1996) Antitumor activity of daphnane-type diterpene gnidimacrin isolated from Stellera chamaejasme L. Int J Cancer 66:268-273

Yu G, Ahmad S, Aquino A, Fairchild CR, Trepel JB, Ohno S, Suzuki K,Tsuruo T, Cowan KH, Glazer RI (1991) Transfection with protein kinase Cα confers increased multidrug resistance to MCF-7 cells expressing P-glycoprotein. Cancer Commun 3:181-189

Yu CW, Chen JH, Lin LY (1997) Metal-induced metallothionein gene expression can be inactivated by protein kinase C inhinbitor. FEBS Lett 420:69-73

Yu L, Orlandi L, Wang P, Orr MS, Senderowicz AM, Sausville EA, Silvestrini R, Watanabe N, Piwnica-Worms H, O'Connor PM (1998) UCN-01 abrogates G2 arrest through a Cdc2-dependent pathway that is associated with inactivation of the Wee1Hu kinase and activation of the Cdc25C phosphatase. J Biol Chem 273:33455-33464

Yuspa SH (1998) The pathogenesis of squamous cell cancer: lessions learned from studies of skin carcinogenesis. J Dermatol Sci 17:1-7

Zellner A, Fetell MR, Bruce JN, De Vivo DC, O'Driscoll KR (1998) Disparity in expression of protein kinase C alpha in human glioma versus glioma-derived primary cell lines: therapeutic implications. Clin Cancer Res 4:1797-1802

Zhang W, Yamada H, Sakai N, Nozawa Y (1993) Sensitization of C6 glioma cells to radiation by staurosporine, a potent protein kinase C inhibitor. J Neurooncol 15:1-7

Zheng B, Chambers TC, Raynor RL, Markham PN, Gebel HM, Vogler WR, Kuo JF (1994) Human leukemia K562 cell mutant (K562/OA200) selected for resistance to okadaic acid (protein phosphatase inhibitor) lacks protein kinase C epsilon, exhibits multidrug resistance phenotype, and expresses drug pump P-glycoprotein. J Biol Chem 269:12332-12338

Zhou T, Song L, Yang P, Wang Z, Lui D, Jope RS (1999) Bisindolylmaleimide VIII facilitates Fas-mediated apoptosis and inhibits T cell-mediated autoimmune diseases. Nature Med 5:42-48

Zhuang S, Lynch MC, Kochevar IE (1998) Activation of protein kinase C is required for protection of cells against apoptosis induced by singlet oxygen. FEBS Lett 437:158-162

Ziegler WH, Parekh DB, Le Good JA, Whelan RD, Kelly JJ, Frech M, Hemmings BA, Parker PJ (1999) Rapamycin-sensitive phosphorylation of PKC on a carboxy-terminal site by an atypical PKC complex. Curr Biol 9:522-529

Zimmermann J, Caravatti G, Mett H, Meyer T, Muller M, Lydon NB, Fabbro D (1996) Phenylamino-pyrimidine (PAP) derivatives: a new class of potent and selective inhibitors of protein kinase C (PKC). Arch Pharm 329:371-376

Zini N, Neri LM, Ognibene A, Scotlandi K, Baldini N, Maraldi NM (1997) Increase of nuclear phosphatidylinositol 4,5-bisphosphate and phospholipase C beta 1 is not associated to variations of protein kinase C in multidrug-resistant Saos-2 cells. Microsc Res Tech 36:172-178

Zunino F, Perego P, Pilotti S, Pratesi G, Supino R, Arcamone F (1997) Role of apoptotic response in cellular resistance to cytotoxic agents. Pharmacol Ther 76:177-
185

Compartment-Specific Functions
of the Ubiquitin-Proteasome Pathway

T. Sommer, E. Jarosch and U. Lenk

Max-Delbrück-Center for Molecular Medicine, Robert-Rössle-Straße 10,
13092 Berlin, Germany

Contents

1 Introduction

Intracellular protein degradation, which was thought to be an unspecific and uncontrolled event, is now known to constitute a tightly regulated process involved in numerous basic cellular functions. The vast majority of specific proteolysis in the cell is executed by the ubiquitin-proteasome pathway. Among the numerous short-lived substrates that have been described, are cell cycle regulators, modulators of transcription, components of signal transduction cascades, enzymes of metabolic pathways, foreign antigens cleaved for antigen presentation but also damaged or mutated proteins. To guarantee specificity, the ubiquitin system is structured in form of a complex cascade of enzymes and recognition factors (Hershko and Ciechanover, 1998; Hochstrasser, 1996). Given the multitude of substrates, it is not surprising that ubiquitin-dependent proteolysis is a key control element in many central cell biological processes. Moreover, aberrations of this system seem to be implicated in the pathogenesis of several important human diseases (Ciechanover, 1998).

Eukaryotic cells represent highly ordered three-dimensional structures in which specific functions are confined to certain areas. In contrast, it was assumed for a long time that the ubiquitin system would be evenly dispersed throughout cytoplasm and nucleus. A number of recent results have changed this view. It is now widely accepted that the ubiquitin-proteasome pathway mediates also the degradation of secretory proteins, which occurs mainly at the endoplasmic reticulum (ER). In this process, malfolded and short-lived secretory proteins are first transported back into the cytosol where they are subjected to ubiquitin-conjugation and proteolysis by the 26S proteasome. It is generally assumed that this proteolytic pathway is restricted to the cytoplasmic surface of the ER (Sommer and Wolf, 1997; Kopito, 1997; Bonifacino and Weissman, 1998). Another example of compartment specific ubiquitin-conjugation takes place at the plasma membrane: ubiquitin-conjugation is an essential step during endocytosis of proteins of the plasma membrane (Hicke, 1999; Strous and Govers, 1999). Moreover, some recent reports have pointed to the fact that degradation of nuclear substrates may depend on the transport of the substrate into or out of the nucleus. In this review we will summarize the current knowledge on the spatial organization of the ubiquitin-proteasome pathway.

2 The Ubiquitin System

Hershko, Rose and colleagues identified the ubiquitin system more than 20 years ago in a cell-free system from rabbit reticulocytes. They demonstrated in vitro that the previously known energy dependent degradation of proteins required the covalent modification of the substrate with the polypeptide ubiquitin. In addition, the enzymological framework leading to ubiquitin-protein conjugates was biochemically dissected, basic enzymes were purified and their activity defined in vitro (Hershko and Ciechanover, 1998). Later, the in vivo relevance of the system was demonstrated first in mammalian cells and then in the yeast *Saccharomyces cerevisiae*. Especially the analysis of yeast mutants in the components of the ubiquitin-conjugation cascade revealed the large variety of processes that are controlled by proteolysis (Hochstrasser, 1996). Thus, a lot of examples described in this review are derived from studies in the yeast system.

Ubiquitin is a highly conserved protein of 76 amino acids, which differs in only three amino acids between yeast and mammals. It is a tightly folded, globular protein with a protruding C-terminus. Determination of the three-dimensional structure revealed that the globular domain exposes patches of different physical properties that probably allow interactions with surfaces of other proteins (Vijay-Kumar, 1985). In all eukaryotic cells, ubiquitin is encoded by three classes of genes that are rapidly processed after translation to yield free ubiquitin. The two classes of genes that provide ubiquitin under normal growth conditions code for fusions to proteins of the large and the small subunit of the ribosome, respectively. The third class encodes head-to-tail fusions of several ubiquitin molecules in which the last unit carries an additional asparagine residue at the C-terminus. These poly-ubiquitin genes are induced under stress conditions. The processing of all these precursors to individual ubiquitin molecules is performed by a group of enzymes called de-ubiquitinating enzymes (see below). Hereby, a Gly-Gly di-peptide is exposed at the C terminus of ubiquitin that is essential for the conjugation to substrate proteins (Jentsch, 1992).

The usual ubiquitin-substrate linkage is an isopeptide bond between internal lysine residues of the substrates and the protruding C-terminus of ubiquitin. However, also the N-terminus of substrate molecules appears to be a possible acceptor for modification with ubiquitin since the ubiquitin-dependent degradation of the transcription factor MyoD proceeded after removal of all the protein's lysine residues (Breitschopf et al. 1998). Ubiquitin itself has no enzymatic activity but when it is covalently linked to other proteins it functions as a tag, marking proteins for destruction by the 26S proteasome. This large protease consists of two types of subparticles: a cylindrical 20S core that harbors the catalytic activities and a 19S regulatory complex attached at either end of the core cylinder (Baumeister et al. 1998). Ubiquitin-conjugated proteins are recognized by the 26S proteasome efficiently, only when multiple ubiquitin molecules are attached to them. Such a poly-ubiquitin chain is formed in successive rounds of conjugation, in which the C-terminus of ubiquitin is linked to a lysine residue of a ubiquitin molecule that has been attached to the substrate in a previous conjugation reaction. The proteolytic relevant ubiquitin-ubiquitin linkage involves Lys48 (Hochstrasser, 1996). Chains linked in such manners create unique hydrophobic patches which are probably recognized by the 19S cap of the proteasome (Pickart, 1997). This entrance control prevents the unspecific degradation of cytosolic proteins. Poly-ubiquitin chains are not static elements, but highly dynamic structures with rapid addition and removal of ubiquitin moieties. Although poly-ubiquitination seems to be the predominant form of conjugation, mono-ubiquitinated species have also been observed in vivo (Hochstrasser, 1996).

In a brief and simplified view, the formation of ubiquitin-conjugates requires the successive action of three classes of enzymes: The E1 or ubiquitin-activating enzyme, E2s or ubiquitin-conjugating enzymes (Ubc) and occasionally E3s or ubiquitin-protein ligases. E1 hydrolyses ATP to first adenylate the C-terminal glycine of ubiquitin and then link it to the side chain of its central cysteine residue, yielding a high energy E1-ubiquitin thioester, free AMP and pyrophosphate. From the E1, the thioester is transferred onto the Ubc's. In vitro, the Ubc's are able to directly link ubiquitin to substrate proteins. However in most cases described in vivo, the conjugation of substrates requires the function of an E3 (Fig. 1; Hershko and Ciechanover, 1998).

The three classes of enzymes build a cascade initiated by the E1 enzyme that is required for all subsequent steps. The second enzymatic activity (E2) comprises a family of enzymes related in sequence. Each of them participates in the turnover of only a limited number of substrates, indicating that these enzymes are among those components that mediate specificity. Given the numerous proteolytic substrates that have to be degraded with different kinetics at any given timepoint in a cell, it is not surprising that the third class of enzymes (E3) constitute a highly diverse group. At least two types can be distinguished by their enzymatic mechanism. One type of E3 enzymes is unable to form transfer intermediates with ubiquitin. However, they possess substrate-binding properties and thus direct the Ubc's to their targets. Representative for this class is the first E3 described (E3α) and its yeast homologue Ubr1 but also the multisubunit E3 complexes (RING-H2-finger E3) which are involved for instance in the proteolysis of cyclins (Deshaies, 1999). Another class of E3s, in addition to substrate recognition, is able to form a DTT-sensitive adduct with ubiquitin (HECT E3s, see below; Scheffner et al. 1995). They are able to directly interact with the substrate or they need ancillary factors that mediate this association (Ciechanover, 1998).

The characteristics of the ubiquitination cascade suggest that the substrate specificity of the ubiquitin system may be extended through association of a limited number of E2 and E3 enzymes into multiple oligomeric complexes (Hochstrasser, 1996).

In addition to this well characterized cascade, a modulator of poly-ubiquitination has been described. This factor, termed E4, promotes the formation of high molecular weight ubiquitin-substrate conjugates in conjunction with E1, E2 and E3. However, E4 is not directly involved in the transfer of ubiquitin to the substrate but binds the ubiquitin-conjugate. In the yeast system, this factor turned out to be identical to Ufd2 (Ubiquitin fusion degradation), which was found previously in a genetic screen. Though it belongs to a family of conserved proteins found in different organisms, it does not perform essential functions in yeast. However, a function under stress conditions can be assumed, possibly by altering the length of the ubiquitin chain (Koegl et al. 1999).

2.1 Ubiquitin-Activating and Ubiquitin-Conjugating Enzymes

The first step in the ubiquitin conjugation cascade is the activation by the E1 enzyme. E1 enzymes and specifically their active site cysteine residue are essential for cell viability (Finley et al. 1984; Ciechanover et al. 1984).

The E2 activity comprises a highly conserved family of enzymes found in all eukaryotic organisms. A domain of about 150 amino acids termed Ubc-domain characterizes them. At a central cysteine residue of this region, the thioester bond with ubiquitin is formed (Jentsch et al. 1990). In addition to the core, Ubc's may carry extensions at the C- or N-terminus or both. These extensions apparently contribute to the substrate specificity of the enzymes. In yeast, where only C-terminal extensions exist, it could be shown that the swapping of the extensions between Ubc3/Cdc34 and Ubc2/Rad6 also switches their substrate specificity (Silver et al. 1992). Furthermore, some of the extensions contain stretches of hydrophobic amino acids integrating the enzymes into a specific cellular membrane (Sommer and Jentsch, 1993).

The analysis of cells lacking a specific E2 demonstrated that they are involved in a large variety of cellular processes, like cell cycle control, DNA repair, peroxisome biogenesis, stress response and resistance to heavy metals. However, the analysis of specific proteolytic substrates revealed that Ubc's seemed to be redundant in function. Furthermore, ubiquitin-conjugating enzymes may function in dimeric complexes with other Ubc's, which may alter the substrate specificity of the involved enzymes (Chen et al. 1993). The current knowledge on the yeast Ubc's is summarized in Table 1.

Table 1. The *Saccharomyces cerevisiae* ubiquitin-conjugating enzymes

Protein	kDa	Functions
Ubc1	24	essential in absence of Ubc4 and Ubc5, important for outgrowth of spores, endocytosis of membrane proteins
Ubc2/Rad6	20	DNA repair, induced mutagenesis, sporulation, repression of retrotransposition, N-end rule pathway
Ubc3/Cdc34	34	mainly nuclear localization, essential for viability, G1-S cell cycle progession, DNA replication, spindle pole body separation, degradation of p40 (Sic1), Gcn4, and Clns
Ubc4 Ubc5	16 16	92% identity between Ubc4 and Ubc5, degradation of abnormal proteins, sporulation, resistance to stress conditions, degradation of MATα2 and ubiquitin-fusion proteins, endocytosis of membrane proteins
Ubc6/Doa2	28	localized at the ER-membrane/nuclear envelope, involved in ERAD, degradation of MATα2
Ubc7	18	localized at the ER-membrane via interaction with Cue1p, central ERAD component, degradation of MATα2, resistance to cadmium
Ubc8	25	catabolite degradation of fructose-1,6-bisphosphatase
Ubc9	18	Smt3 conjugating enzyme, mainly nuclear localization, essential for viability, mutants arrest at G2-M transition of the cell cycle, deficient in degradation of Clb2 and Clb5 cyclins
Ubc10/Pas2	21	peroxisome biogenesis
Ubc11	17	not essential also in combination with Ubc4
Ubc12	21	Rub1 conjugating enzyme
Ubc13	18	DNA repair, modified by Mms2 to form K63 linked polyubiquitin chains

2.2 Ubiquitin-Ligases

In vivo participation of E3 enzymes can be assumed generally and it seems likely that proteolysis is controlled primarily by regulating the activity of the E3's. Although they are thought to be most directly involved in substrate recognition this class of enzymes is least well understood. E3s might be divided into distinct families with regard to sequence class and enzymatic

mechanisms. The first ubiquitin ligase characterized was E3α of rabbit reticulocytes. It is the likely counterpart of yeast Ubr1, a protein of 225 kD.
Ubr1 is the recognition component of the so-called N-end rule pathway
(Bartel et al. 1990). Both enzymes stimulate substrate degradation, and bind
both the substrate and the respective Ubc. In Ubr1 a RING-H2 finger domain is found which is critical for the formation of a poly-ubiquitin chain on
N-end rule substrates (Xie and Varshavsky, 1999).

An important group of E3s is characterized by a conserved element, the
HECT-domain. HECT-domain proteins carry a C-terminal region of about
100 amino acids exhibiting strong similarities to human E6 associated protein (E6AP; Huibregtse et al. 1995). Cellular E6AP together with an ancillary
factor, the papilloma virus E6 oncoprotein, is needed for the turnover of the
tumor suppressor p53 (Scheffner et al. 1993). E6AP is able to form a
thioester bond with ubiquitin at a cysteine residue within the C-terminal
conserved region (Scheffner et al. 1995). It receives the ubiquitin moiety
from a specific Ubc. Thus, E6AP is integrated into the transfer cascade for
ubiquitin. Because of this, a stable interaction of E6AP with a certain Ubc
might not be necessary.

A major input into the architecture and function of E3's were derived
from studies of the yeast cell cycle. In an in vitro reconstituted system the
multimeric ubiquitin ligase SCF was identified. The minimal unit consists of
four different subunits: Cdc53, Hrt1, Skp1 and Cdc34. Cdc53 and Hrt1 recruit the ubiquitin-conjugating enzyme Cdc34 into the complex. Skp1 helps
to link the minimal SCF via interaction with Cdc53 to a member of the family of F-box proteins. The F-box proteins are characterized by the F-box that
binds to Skp1. In addition, they comprise other protein-protein interaction
domains that specifically recognize phosphorylated substrates. Thus, the
different F-box proteins integrate SCF into substrate specific degradation
pathways. Hrt1 also belongs to a family of proteins that are characterized by
a zinc-binding RING-H2 finger domain. Hrt1 dramatically increases the
activity of SCF and enables the complex to synthesize unanchored poly-
ubiquitin chains and promotes auto-ubiquitination of Cdc34. Cdc53 contains a Cullin domain, which is also found in many other proteins. Some of
these protein motives are also found in another ubiquitin ligase complex
that functions at the end of mitosis, the APC complex. Apc2 belongs to the
family of Cullins and Apc11 contains a RING-H2 finger domain (Deshaies,
1999).

2.3 Ubiquitin Specific Proteases

Most of the work of the past years has focused on the enzymes attaching ubiquitin to substrate proteins. Recent results suggest that regulatory events also occur at the level of de-ubiquitination. De-ubiquitinating enzymes are thiol proteases which can be divided into two classes: The so-called ubiquitin-specific processing proteases (Ubp) and the ubiquitin carboxyl-terminal processing hydrolases (Uch) (Wilkinson, 1997). The first class is extremely divergent but all members contain short conserved sequence motifs, termed the Cys and His boxes, respectively. These motifs are likely to form parts of the active site of the enzymes (Hochstrasser, 1996). Two representatives of this enzyme family have been investigated more closely. One of them is the yeast Ubp4/Doa4, a de-ubiquitinating enzyme cleaving linear ubiquitin-protein fusions and isopeptide linkages. In *doa4* mutants ubiquitin-dependent protein degradation of certain N-end-rule substrates is affected (Papa and Hochstrasser, 1993). In addition, ubiquitin species accumulate that are slightly larger than free ubiquitin, suggesting that they represent ubiquitinated peptides that are proteolytic remnants of the 26S proteasome activity. Furthermore, the proteolysis of ubiquitin is increased in *doa4* mutants, suggesting that Doa4 is required to recycle ubiquitin from proteasome-bound substrates (Swaminathan et al. 1999). In agreement with this hypothesis, Ubp4/Doa4 could be co-purified with the 26S proteasome (Papa et al. 1999). Interestingly, Ubp4/Doa4 exhibits sequence similarities to the mammalian tre-2 oncogene and it could be shown that tre-2 is indeed an Ubp (Papa and Hochstrasser, 1993).

The second investigated enzyme of this class is yeast Ubp14, a functional homologue of human isopeptidase T. Cells lacking this activity exhibit no obvious growth defect but are hypersensitive to canavanine, an amino acid analog which leads to aberrant translation products. Furthermore, the turn-over of some typical substrates of the ubiquitin system is reduced. Cells lacking Ubp14 accumulate unanchored poly-ubiquitin chains. These ubiquitin chains do not only derive from the proteolysis of ubiquitinated substrates. Instead cells seem to synthesize unanchored ubiquitin chains de novo. Accumulation of these chains in *ubp14* cells inhibits degradation probably by competing for the substrate-binding sites of the proteasome. Overexpression of Ubp14 and thus reducing free ubiquitin chains also reduces protein breakdown indicating that the pre-assembled ubiquitin chains contribute significantly to the efficiency of ubiquitination (Amerik et al. 1997).

2.4 Ubiquitin-Like Proteins

One of the major developments of the last years was the identification of proteins that display similarities to ubiquitin (Ubl, ubiquitin-like proteins). Like ubiquitin, some of them are covalently linked to a number of cellular proteins and the enzymatic activities required for this modification parallel those linking ubiquitin to substrates. Because the Ubl's have been described only recently, their function is less clear. From the current knowledge it can be speculated that Ubl-conjugation is not directly linked to proteolysis. Instead, at least one of them has been implicated in protein localization. Ubl proteins are less conserved than ubiquitin among species. One of them is the 11.5 kD mammalian protein SUMO-1 which shares 18% sequence identity with ubiquitin. In yeast a homologous protein exists, Smt3, sharing 48% identical amino acids with SUMO-1. Many amino acids involved in interaction of ubiquitin with the proteasome are not present in SUMO-1 and Smt3, respectively. Nevertheless, both can be found either free or conjugated to cellular proteins. Cells lacking Smt3 are inviable, and thus it can be assumed that this Ubl fulfills essential cellular functions.

Another Ubl is NEDD8 and its yeast equivalent Rub1 (59% sequence identity). Unlike Smt3, Rub1 is not essential for viability. NEDD8 displays 58% sequence identity with ubiquitin and 20% with SUMO-1. Two observations suggest that Rub1 is conjugated analogously to ubiquitin: First, identical residues between ubiquitin and Rub1 are clustered on one surface of the three-dimensional structure of ubiquitin and particularly at the C terminus through which ubiquitin is conjugated. Second, Lys48 and Lys29 of ubiquitin are conserved in Rub1. Both residues are required for the formation of a poly-ubiquitin chain. Indeed it turned out that Rub1 is also conjugated to a small number of cellular substrates. Interestingly, one of the substrates is Cdc53, a protein functioning in ubiquitin-protein ligation (Hochstrasser, 1998).

Ubl's are activated by distinct E1 activities but this activity is not provided by a single polypeptide chain. Instead, a heterodimeric enzyme performs the Smt3 or Rub1 activation. The two subunits of these dimers exhibit homologies to either the N- or C-terminal part of the ubiquitin-activating enzyme. In the case of yeast Smt3, Uba2 and Aos1 provide the activating activity (Johnson et al. 1997). Another activating enzyme is formed by Uba3 and Ula1/Enr2 that activates Rub1 (Liakopoulos et al. 1998). A similar activity has been purified from Arabidopsis indicating a high degree of conservation. It comprises Axr1 and Ecr1 that are similar to Ula1/Enr2 and Uba3 respectively. When combined they are able to activate NEDD8 in vitro through thioester formation at Ecr1. Both Ubl's are also conjugated by distinct ubiquitin-conjugating enzymes. Ubc9 conjugates Smt3 (Schwarz et al. 1998) while Ubc12 links Rub1 to substrates (Liakopoulos et al. 1998). In vitro, both Ubc9 and Ubc12 are unable to form thioester bonds with ubiquitin. Ubiquitin ligases specific for Smt3 and Rub1 have not been described yet.

Furthermore, Ubl-specific proteases have been identified in yeast (Li and Hochstrasser, 1999; Li and Hochstrasser, 2000) and also from higher eukaryotic cells (Gong et al. 2000). Yeast Ulp1 and Ulp2 (Ubl-specific protease) cleave Smt3, but not ubiquitin, from substrates. Both enzymes shows no

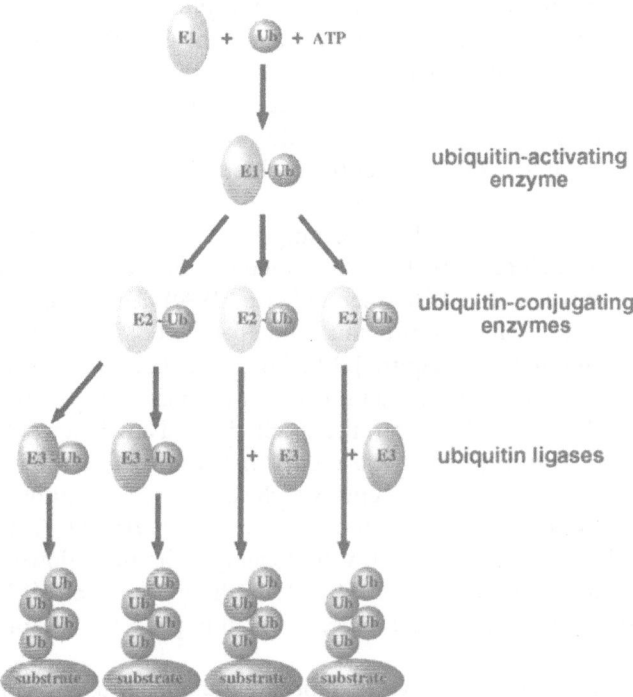

Fig. 1. A schematic diagram outlining the hierarchic structure of the ubiquitin system. In an ATP-dependent manner a thioester bond is formed between the C-terminus of ubiquitin and an internal cystein residue of the ubiquitin-activating enzyme. Subsequently, ubiquitin is transferred to a member of the family of ubiquitin-conjugating enzymes, which are also able to form a thioester bond with ubiquitin. The third class of enzymes, the ubiquitin ligases, direct ubiquitin to the proteolytic substrates. Different families of this class of enzymes are known, some of which are also able to form a thioester intermediate with ubiquitin (HECT-domain ligases). The final ubiquitin-substrate linkage is an isopeptide bond between the C-terminus of ubiquitin and internal lysine residues in the substrate proteins

homology to any known de-ubiquitinating enzymes. Intriguingly, Ulp1 plays an essential role at the G2/M transition of the cell cycle (Li and Hochstrasser, 1999).

3 Retrograde Transport From the ER

3.1 ER Proteins Are Degraded by the Ubiquitin-Proteasome System

The endoplasmic reticulum (ER) constitutes the site of synthesis and maturation of proteins destined for the secretory pathway. Proteins that enter this pathway are inserted into the ER as nascent polypeptide chains. Subsequently, these proteins adopt their proper folding within the ER and are in some cases assembled into multimeric protein complexes, which are then transported to other organelles. Misfolded or aberrantly assembled polypeptide chains are recognized by a specific quality control system that retains such proteins within the ER and eventually channels them to degradation. The importance of ER protein degradation for cellular processes is emphasized by severe diseases, which are accompanied with the breakdown of misfolded proteins, such as cystic fibrosis. Proteolysis of ER proteins has been shown to be highly selective (Klausner and Sitia, 1990). Additionally, it was observed that degradation of ER proteins is independent of lysosomal or vaccuolar proteases and does not require ER to Golgi transport. Thus it was long believed that breakdown of ER proteins occurs within the compartment itself (Klausner and Sitia, 1990). However, the specificity and efficiency of ER degradation was difficult to explain with a model that would place the vast amount of loosely and partly folded proteins, which should be susceptible to proteolysis, and highly active proteases into the same compartment. At the same time, ER localized proteases, which would catalyze the breakdown of substrate proteins, could not be isolated. Therefore, the mechanisms of ER protein degradation and the proteases involved in this process remained mysterious.

In 1993 the identification of a yeast ubiquitin conjugating enzyme, Ubc6p, as an integral protein of the ER membrane with the catalytic part facing the cytosol was reported (Sommer and Jentsch, 1993). Based on the observation that a mutation within the Sec61 protein, which causes a translocation defect, was suppressed by loss of function mutants of Ubc6p the authors proposed for the first time a link between ER protein degradation and cytosolic ubiquitin-proteasome mediated proteolysis (Sommer and Jentsch, 1993). Another line of evidence for an involvement of the ubiquitin-proteasome machinery in the breakdown of ER proteins came from the investigation of the turnover of the cystic fibrosis transmembrane conductance regulator (CFTR). CFTR constitutes a chloride channel of the ATP-binding-cassette class protein family in epithelial cells. Wild type CFTR gets transported to the plasma membrane whereas a mutant form CFTR ΔF508 is retained within the ER and rapidly degraded (Kopito, 1999). Ward et al. and Jensen et al. noticed that treatment of cells expressing CFTR ΔF508 with specific proteasome inhibitors resulted in a remarkable stabilization of the protein (Jensen et al. 1995; Ward et al. 1995). The overexpression of a mutant form of ubiquitin UbK48R, which fails to form poly-ubiquitin chains and thus does not efficiently label substrate proteins for recognition by the 26S proteasome, also resulted in a reduced turnover of CFTR ΔF508. Finally, in the presence of the proteasome inhibitor lactacysteine CFTR accumulated with covalently attached poly-ubiquitin chains (Ward et al. 1995). Recent findings indicate, that the ubiquitin-proteasome machinery also contributes to the breakdown of soluble substrate proteins in the ER lumen (Hiller et al. 1996; Qu et al. 1996; Bonifacino and Weissman, 1998). It is worthy to note that not only misfolded and aberrant proteins are degraded by this mechanism. Proteasome mediated ER protein degradation also has a regulatory function, as has been demonstrated for the catabolite induced degradation of HMG-CoA reductase (Hampton et al. 1996), and accounts for the turnover of otherwise stable host-proteins after viral infections (Wiertz et al. 1996a,b).

Although studies exist which report an association of the 26S proteasome with the ER membrane, proteasomes have never been found inside the ER. It was therefore hard to envisage, how a cytosolic multienzyme protease like the 26S proteasome should catalyze the proteolysis of ER proteins, which are integrated into a membrane by the virtue of several transmembrane segments or are located within the ER lumen and thus separated from the proteasome by a lipid bilayer. Consequently, one can propose certain steps, which an ER degradation substrate should pass in order to become accessible by a cytosolic protease (for a schematic view see Fig. 2). Proteolysis of aberrant ER proteins should be initiated by the recognition of the substrate by the quality control system within the ER. Genetic and biochemical evidence implicates chaperones and folding enzymes as well as components of the glycosylation machinery with this process (Ellgaard et al. 1999). Subsequently, a mechanism should exist that shuttles luminal substrate proteins through the ER membrane into the cytosol. Integral ER membrane proteins must get dislocated from this membrane to become fully accessible by the 26S proteasome. Although we lack a detailed knowledge on these processes, recent studies indicate that the machineries mediating protein import into the ER and protein dislocation from the ER share at least some components, such as the Sec61 translocation channel (Wiertz et al. 1996b; Plemper et al.

1997). Protein translocation through biological membranes depends on targeting signals, which are recognized by specific receptors and initiate the insertion of the substrate into a transport channel. In the case of protein translocation into the ER these signals are often proteolytically removed from the substrates during import (Rapoport et al. 1996). To date, we have no knowledge on the nature of such signals in protein export from the ER, however, it is obvious that import into the ER and dislocation from the ER must depend on different signals. Active transport of proteins through membranes also relies on a driving force, which determines the direction of protein movement. No such driving force has been described in retrograde transport, so far, although ER protein degradation depends on several components, which hydrolyze nucleotide-tri-phosphates during their enzymatic activity. Once in the cytosol the substrates are labeled by the covalent attachment of poly-ubiquitin and are finally degraded by the 26S proteasome. In the following chapters we will try to give an overview on the current knowledge on proteasome mediated degradation of ER proteins and summarize the implications resulting from the models proposed for the mode of action of the ubiquitin-proteasome machinery on ER substrate proteins.

3.2 Systems to Study ER Protein Degradation

3.2.1 Yeast

Genetic and biochemical studies on the breakdown of a variety of substrate proteins in the yeast *Saccharomyces cerevisiae* have resulted in major contributions for the understanding ER protein degradation (Brodsky and McCracken, 1997; Sommer and Wolf, 1997; Bonifacino and Weissman, 1998; Plemper and Wolf, 1999). Sec61, a multispanning ER membrane protein, constitutes an essential component of the Sec61 complex that mediates translocation of proteins into the ER. A mutation in Sec61, termed sec61-2, results in an unstable protein which causes the dissociation of the Sec61 complex and a translocation defect (Sommer and Jentsch, 1993; Biederer et al. 1996). Degradation of Sec61-2 was shown to depend on poly-ubiquitination by the E2 enzymes Ubc6 and Ubc7 and on proteasomal activity (Biederer et al. 1996). Similarly, a mutant of the multidrug-resistance-mediating protein Pdr5, an ATP binding cassette type plasma membrane channel that is integrated into the membrane by the virtue of 12 transmembrane segments, undergoes rapid turnover in the ER. The degradation of Pdr5* parallels the breakdown of Sec61-2 in it's need for ubiquitination and proteasome activity (Plemper et al. 1998). The turnover of a mutant form of uracil permease, which normally constitutes an integral protein of the plasma membrane, has likewise been shown to occur in an ubiquitin-proteasome dependent manner associated with the ER (Galan et al. 1998). Degradation by the 26S proteasome does not only affect integral membrane proteins, but also accounts for the proteolysis of soluble substrates in the ER lumen. CPY*, a mutant allele of vacuolar carboxypeptidase Y, is retarded in the ER lumen and rapidly degraded (Finger et al. 1993; Hiller et al. 1996). Proteolytic breakdown depends on proteasomal functions, as well as ubiquitin-conjugation and chaperones within the ER lumen (see below). Another soluble substrate in the ER lumen is an unglycosylated mutant of yeast α-factor. Haploid yeast cells exist in two different mating types: a and α. A

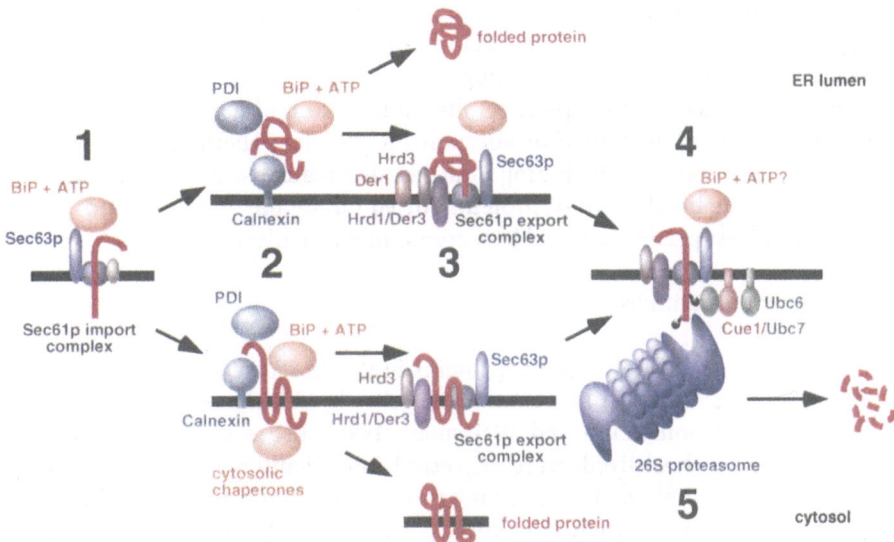

Fig. 2. Schematic model proposed for the degradation ER proteins. The protein gets translocated into the ER by the Sec61 translocation machinery (1) and subsequently associates with luminal chaperones and folding enzymes (2). Association with BiP, PDI, calnexin and glycosylating enzymes, which are part of the quality control system (see text), will either result in folding and release of the enzyme for further transport or target the substrate to degradation. The folding of cytoplasmic domains present in integral membrane proteins requires the action of cytosolic chaperones. Extraction from the membrane and retrograde transport from the ER (3) involves the Sec61 translocon and probably additional components, like the yeast Hrd and Der proteins (see text). At the cytosolic side, poly-ubiquitin is covalently attached to the substrate by E2 enzymes located at the membrane (4), which labels proteins for degradation and most likely contributes to the transport process (see text). Finally, substrate breakdown is mediated by cytosolic 26S proteasomes (5)

cells secrete a small glycosylated protein, termed α pheromone or α-factor, into the medium which binds to the pheromone receptors of a cells and causes morphological changes required for mating. By removal of the glycosylation sites pre-pro-α-factor becomes an unstable protein that undergoes ER protein degradation (Caplan et al. 1991). Yeast pre-pro-α-factor has served as a substrate protein in a reconstituted system to study ER protein degradation in vitro (see later).

Physiological regulation of cellular processes by controlled protein degradation by the 26S proteasome has been observed in several cases (Hochstrasser, 1995; King et al. 1996; Hershko, 1997). 3-hydroxy3-methylglutaryl-CoA reductase (HMG-CoA reductase), an integral protein of the ER membrane involved in the mevalonate pathway, is subjected to feedback regulation and undergoes regulated degradation via the ubiquitin-proteasome pathway in mammals and yeast (Hampton and Rine, 1994; Hampton et al. 1996; McGee et al. 1996). Interestingly, degradation of the yeast HMG-CoA reductase was shown to depend on the same factors which are required for the proteolysis of malfolded or aberrant ER proteins (Hampton et al. 1996; Bordallo et al. 1998). This finding indicates that regulated ER protein degradation and the breakdown of proteins sorted out by the quality control system are at least partly mediated by a common mechanism.

3.2.2 MammalianCells

The turnover of several unstable proteins in the secretory pathway has been intensively examined in mammalian cells by biochemical techniques (Kopito, 1997; Bonifacino and Weissman, 1998). Proteins, that have been reported to be short-lived, were expressed in cell-cultures to allow the detailed study of their maturation and proteolysis. Cystic fibrosis, a human autosomal recessive genetic disorder, is caused by the dysfunction of the cystic fibrosis transmembrane conductance regulator (CFTR), an ATP-binding cassette-type chloride channel in the plasma membrane (Kopito, 1999). Mutant ΔF508 of CFTR is of special interest: The majority of patients suffering from cystic fibrosis carry the CFTR ΔF508 allele. CFTR ΔF508 does not get transported to the cell surface but is retarded in the ER and rapidly degraded in a ubiquitin-proteasome dependent manner (Jensen et al. 1995; Ward et al. 1995). Overexpression of CFTR ΔF508 in transfected cells results in the appearance of functional channel molecules in the plasma membrane indicating that the pathogenesis of cystic fibrosis is most likely caused primarily from mislocalization and rapid degradation of the CFTR ΔF508 mutant protein (Dalemans et al. 1991). Indeed, CFTR ΔF508 is transported to the cell surface and forms active Cl-channels, when certain arginine residues present in the luminal part of protein were changed to lysines thereby masking a potential ER retardation signal and allowing the protein to escape ER protein degradation (Chang et al. 1999). Surprisingly, maturation and transport of the wild-type CFTR is also ineffective in transfected cells indicating the requirement for additional factors for folding and membrane delivery that may be limiting in these cells (Lukacs et al. 1994; Ward and Kopito, 1994). A mutant in α1-antitrypsin (α1-ATZ), that causes retardation of this soluble secretory protein within the ER lumen, has been shown to account for chronic liver injuries and hepatocellular carcinoma. Biochemical investigation of α1-ATZ revealed, that it is degraded in an ubiquitin-proteasome dependent manner (Qu et al. 1996).

In the ER proteins are not only folded into their mature structure but are in some cases integrated into multimeric enzyme complexes. Components of such complexes, which are present in excess and will therefore not be correctly assembled, are removed from the ER and degraded. A model substrate for this process is the T-cell antigen receptor, which represents a hetero-oligomeric plasma membrane complex consisting of at least seven transmembrane subunits (Klausner and Sitia, 1990). In the absence of other subunits, the TCRα, an integral membrane protein, is retarded within the ER and rapidly degraded. A detailed analysis of TCRα proteolysis revealed that it paralleled the breakdown of CFTR ΔF508 in the requirement for poly-ubiquitination and proteasomal function, thus representing another example of ubiquitin-proteasome mediated ER protein degradation (Huppa and Ploegh, 1997; Yang et al. 1998; Yu and Kopito, 1999).

3.2.3 Viral Induction of Proteolysis

MHC class I restricted antigen presentation provides a potential of the cellular immune system to selectively label virus infected cells at the surface and kill them by CD8+ T-lymphocytes. This mechanism requires that MHC class I molecules are loaded with virus derived peptides in the ER and transported by the secretory pathway to the plasma membrane. The human cytomegalo virus (HCMV) employs a strategy to interfere with this process and escape detection by the immune system. HCMV encodes two proteins, US2 and US11, which interact with MHC class I molecules in the ER membrane and target them to proteasome mediated degradation (Wiertz et al. 1996a). This process depends on the ubiquitination of the substrate protein as well as proteasomal functions. Breakdown of ER proteins by the expression of viral gene-products features several mechanistic differences when compared to the extraction and degradation of ER substrates sorted out by the quality control. In US2 expressing cells de-glycosylated MHC class I molecules have been shown to associate transiently with Sec61β, whereas the proteolysis of MHC class I molecules due to misfolding of the substrate, e.g. after treatment with DTT, involves a Sec61β associated and glycosylated degradation intermediate (Wiertz et al. 1996b). Furthermore, membrane-extraction of MHC class I molecules in US11 expressing cells has been reported to be initiated independently from poly-ubiquitination (Shamu et al. 1999). On the contrary, attachment of ubiquitin constitutes a pre-requisite for the dislocation of other degradation substrates in mammalian and yeast cells, such as TCRα and CPY* (Biederer et al. 1997; Yu and Kopito, 1999).

3.2.4 In Vitro Systems

McCracken and Brodsky have established a reconstituted system to examine the turnover of an ER protein in vitro (McCracken and Brodsky, 1996). Radio-labeled mutant pre-pro-α-factor, which has been depleted of all glycosylation sites and thus becomes an unstable protein of the ER lumen (see above), was in vitro translocated into microsomes isolated from yeast cells. After import and removal of substrate molecules that did not get translocated into the vesicles addition of cytosol and an ATP-regenerating system resulted in the export and degradation of pre-pro-α-factor. Proteolysis of pre-pro-α-factor in this system depended on proteasomal functions, as cytosol derived from yeast proteasomal mutants failed to promote degradation. However, breakdown of mutant pre-pro-α-factor in vitro most likely represents a special case of ER protein degradation. The removal of glycosylation sites results in de-stabilization of the pro-α-factor, whereas such mutations in other substrates cause accumulation within the ER (Knop et al. 1996a). Moreover, turnover of pre-pro-α-factor has been shown to occur independently from poly-ubiquitination and the involvement of other known components of the ER protein degradation machinery, like the Hrd and Der proteins described below, has not been demonstrated, yet. Similar in vitro systems to study the degradation α1-ATZ and CFTR ΔF508, respectively, have been recently reported (Qu et al. 1996; Sato et al. 1998; Xiong et al. 1999).

3.3 The ER Quality Control System

Not much is known about the mechanisms that ensure efficient discrimination between ER proteins, which are in the process of folding and will finally form stable enzymes, and those, which fail to acquire their mature structure and have to be degraded. For example, small peptide stretches, which are buried in the mature but exposed in a native conformation of a protein, have been proposed to serve as an indicator for the correct folding and maturation state of a substrate. Yet, such stretches are likely to be exposed also in intermediates of the folding process of functional enzymes that have not yet acquired their final structure. Therefore, breakdown of a substrate protein must be delayed for the time needed to become at least partly folded. Newly translocated proteins associate with chaperones and folding enzymes within the ER lumen such as BiP, calnexin, calreticulin, the protein disulfide isomerase, GRP94, ERp57 and ERp72 (Ellgaard et al. 1999). Prolonged association with these folding enzymes due to an inability to fold the substrates appropriately may constitute one mechanism to sort out proteins for degradation (Knittler et al. 1995; Ellgaard et al. 1999). Indeed, the involvement of ER localized chaperones for the efficient turnover of ER degradation substrates has recently been demonstrated. Microsomes harboring mutant versions of calnexin (McCracken and Brodsky, 1996) or Kar2 (Brodsky et al. 1999), the yeast homologue of BiP, were unable to induce efficient export and proteolysis of pre-pro-α-factor in vitro. Degradation of α1-ATZ in vivo involves the association of the substrate with ubiquitinated calnexin, suggesting role of this chaperone in recognition and targeting of ER degradation substrates to proteolysis (Qu et al. 1996). A function of BiP in ER protein degradation was genetically demonstrated in the yeast *Saccharomyces cerevisiae*. Mutants of BiP failed to degrade the substrate protein CPY* under conditions that did not affect import of this protein into the ER (Plemper et al. 1997). The activity of ER chaperones and folding enzymes depends on the Ca^{2+} concentration within the ER lumen. In yeast, disruption of the ion transporter Pmr1, which supplies the ER with the divalent cations Ca^{2+} and Mn^{2+}, has been shown to cause stabilization of the ER degradation substrate CPY* (Dürr et al. 1998). However, the effects of a *pmr1* deletion on ER protein degradation are most likely pleiotropic because changes in the ionic milieu within the ER lumen probably affect chaperone activity as well as other enzymatic reactions such as glycosylation. Recently, degradation of mutated pre-pro-α-factor in vitro was shown to require the function of protein disulfide isomerase (PDI) (Gillece et al. 1999). The retrograde transport of pre-pro-α-factor depended on the interaction with PDI, because microsomes harboring different deletion mutants of PDI were unable to export this substrate. Interestingly, export and proteolysis of pre-pro-α-factor was not affected by a point mutation within PDI, which disrupts only the catalytic activity, but still allowes binding of PDI to substrate proteins. Controversingly, degradation of CPY* was hardly affected in the PDI deletion mutants, however, CPY* breakdown still depended on disulfide bond formation (Gillece et al. 1999). Protein folding and disulfide bond formation may thus represent distinct enzymatic functions of PDI. The exact role of chaperones in ER protein degradation has to be determined further. It was speculated that BiP or other ER folding enzymes contribute to the dislocation of ER degradation substrates by constituting a motor to push substrate proteins out of the ER and into the cytosol (see later). There exists no biochemical evidence for such a mode of action, so far. More likely, these enzymes may primarily fulfill other functions such as preventing the aggregation of aberrant proteins and keeping them in a dislocation competent conformation, or the targeting of substrates to an export channel. BiP may also act as a gating factor in protein import and protein dislocation that seals the translocation pore to prevent the uncontrolled diffusion of small molecules trough the Sec61 channel (Hamman et al. 1997). The capability of BiP to bind to nascent chains in yeast depends on the interaction with Sec63, an essential protein required for post-translational import into the ER (Scidmore et al. 1993; Rapoport et al. 1996). Interestingly, mutants in Sec63 have also been shown to stabilize ER degradation substrates in yeast, suggesting a function of this enzyme in the turnover of ER proteins (Plemper et al. 1997). It is not clear, however, whether Sec63 and BiP interact during ER protein degradation.

The correct folding of the cytoplasmic domains of integral ER membrane proteins requires the action of cytosolic chaperones. There exist controversial reports about an involvement of these enzymes in ER protein degradation. For example, degradation of an unstable mutant form of the insulin receptor was found to depend on cytosolic Hsp90 (Imamura et al. 1998). Injection of specific anti-Hsp90 antibodies into cells resulted in a remarkable stabilization of the receptor. On the contrary, disruption of the binding of cytosolic Hsp90 to CFTR by the drug Geldanamycin resulted in an accelerated breakdown of CFTR as well as the unstable mutant CFTR ΔF508 (Loo et al. 1998). Cytosolic Hsp70 and Hsp90 have also been found in association with cytoplasmic aggregates of CFTR ΔF508 in cells treated with proteasome inhibitors. Still, these Hsps probably did not promote dislocation of CFTR but rather associated with the substrate afterwards. A function of these chaperones in the degradation of luminal ER proteins has not been reported, so far, questioning a general role for these proteins in the dislocation of ER substrates.

3.3.1 Glycosylation

Certain proteins destined for the plasma membrane or other organelles are modified by the covalent attachment of sugar residues within the ER. This glycosylation represents a dynamic structure, which is repeatedly trimmed and re-synthesized by a multitude of enzymes. Surprisingly, the inhibition of mannose trimming in the ER results in the stabilization of otherwise unstable glycoproteins which indicates that only glycoproteins containing completely maturated mannose residues will be accepted as substrates of the degradation machinery. (Su et al. 1993; Helenius, 1994; Knop et al. 1996a). The degradation of unassembled subunits of the asialoglycoprotein receptor is severely affected by inhibiting the trimming of mannose chains with chemicals (Ayalon Soffer et al. 1999). Binding of the substrate to calnexin was not affected under these conditions. This suggests that the mannosidases localized in the ER may serve as sensors for the folding state of a glycoprotein. In yeast, mutations, that prevent the mannose trimming catalyzed by mannosidase I impair the proteolysis of CPY* (Knop et al. 1996a; Jakob et al. 1998). Strikingly, the processing of glycan-residues present in CPY* in the ER takes about 20 min., which is in the range of the observed half live (Jakob et al. 1998). This finding suggests that mannose trimming may cause a delay in the action of the quality control system which is required for the correct folding of proteins by the folding machinery and allows functional enzymes to acquire their mature structure before degradation is initiated. Similar results have been obtained in mammalian cells by the investigation of mannose trimming during the breakdown of mutant ribophorin I (de Virgilio et al. 1999). Inhibition of glucose chain trimming in mammalian cells has the opposite effect: Applying the drug Castanospermine to cells increases the degradation of substrate proteins as has been demonstrated for unassembled subunits of the asialoglycoprotein receptor (Ayalon Soffer et al. 1999) and the α-subunit of the nicotinic acetylcholine receptor (Keller et al. 1998).

3.4 Specific Factors Involved in ER Protein Degradation

The search for yeast mutants, which were defective in the turnover of HMG-CoA reductase and CPY*, respectively, resulted in the identification of several genes, whose function had not been described at that time. (Hampton et al. 1996; Hiller et al. 1996; Knop et al. 1996b). Surprisingly, both genetic screens resulted in the isolation of an overlapping spectrum of genes which indicated the existence of a single machinery required for the breakdown of misfolded proteins sorted out by the quality control system as well as substrates that are degraded in a regulated manner. Disruption of these newly isolated genes did not cause a growth defect or other detectable phenotype, suggesting that ER protein degradation is not essential for the viability of yeast cells. *HRD1/DER3* encodes an integral protein of the ER membrane with a large luminal domain (Bordallo et al. 1998). Interestingly, this domain harbors a region with high similarity to the so-called RING-H2 finger motif, which is also found in a subunit of the E3-SCF complex. The RING-H2 motif of Hrd1/Der3 is essential for its function because deletion of this region or a point mutation changing a single cysteine residue to serine results in a biologically inactive protein (Bordallo et al. 1998; Bordallo and Wolf, 1999). This led to the speculation that Hrd1 may be a component of an E3 enzyme, which catalyzes the poly-ubiquitination of ER degradation substrates. All features found for Hrd1/Der3 point to such a function. The RING-H2 domain in the SCF complex has been demonstrated to mediate the association with E2 enzymes. Whether the Hrd1/Der3 RING-H2 motif is able to interact with cytosolic E2 enzymes has still to be shown. Meanwhile, the Hrd1/Der3 protein has shown to be necessary for the proteolysis of all yeast ER degradation substrates investigated so far, which points to a central function of Hrd1/Der3 in ER protein turnover. Hrd3 constitutes an integral protein of the ER membrane with one transmembrane segment. Surprisingly, deletion of *HRD3* results in stabilization of HMG-CoA reductase and CPY* but rapidly increases the turnover of Hrd1/Der3. Accelerated degradation of Hrd1 in *HRD3* disrupted cells depends on proteasomal activity as well as Ubc7 and requires Sec61 function (Plemper et al. 1999b). It was therefore speculated that Hrd1 and Hrd3 form a protein complex in the ER membrane, which dissociates in the absence of Hrd3 and leads to Hrd1 degradation. Another integral membrane protein, which is required for the degradation of CPY*, is Der1 (Knop et al. 1996b). Surprisingly, Der1 function is dispensable for the breakdown of other ER degradation substrates tested so far (Bordallo et al. 1998) but constitutes a necessity for the rapid turnover of Hrd1/Der3 in *HRD3* deletion strains (Plemper et al. 1999b). The function of Hrd3 and Der1 in ER protein degradation has not been established yet. These two proteins together with Hrd1/Der3 have been suggested to associate in a complex, which promotes the dislocation and proteolysis of ER degradation substrates. The luminal domain of Hrd1/Der3 may also be involved in the targeting of substrate proteins to the translocation pore. Further biochemical studies will be needed to determine the function of Hrd1, Hrd3 and Der1 in ER protein turnover.

Two ubiquitin conjugating enzymes (E2s) have been shown to contribute to the degradation of ER protein substrates in yeast (Sommer and Jentsch, 1993; Biederer et al. 1996; Hiller et al. 1996). One of these, Ubc6, is integrated into the ER membrane by a single transmembrane segment, the catalytic domain facing the cytosol (Sommer and Jentsch, 1993). The other one, Ubc7, constitutes a soluble protein, which is recruited to the ER membrane by the interaction with the integral membrane protein Cue1 (Biederer et al. 1997). Both enzymes have been shown to be also required for the breakdown of soluble non-ER proteins (Chen et al. 1993). The interaction with Cue1 is essential for Ubc7 function: In a Cue1 deletion strain Ubc7 is mislocalized to the cytosol and degraded (Biederer et al. 1997). As a consequence, substrate proteins, which are degraded in an Ubc7 dependent manner, are stabilized. Interestingly, deletion of Cue1 does not only affect the breakdown of ER degradation substrates, but also results in the stabilization of non-ER proteins. Overexpression of Ubc7 in the absence of Cue1 to obtain wild type protein levels does not overcome the loss of Cue1. This finding indicates that localization of Ubc7 at the ER membrane is an essential prerequisite for its function, emphasizing the importance of correct cellular placement for enzymatic activities. Disruption of either *UBC6* or *UBC7* affects the breakdown of all yeast ER degradation substrates tested so far, stressing the important role of these E2s in ER protein turnover.

To date, no mammalian factors specifically involved in ER protein degradation have been isolated. There exist entries in the sequence databases, which share similarity to the yeast Hrd1/Der3, Hrd3 and Der1 proteins, respectively, however, the function of these proteins has not been determined so far. Therefore, it remains an open question whether ER protein degradation in mammalian cells requires the activity of similar cellular components that have been shown to mediate turnover of ER proteins in yeast.

3.5 Retrograde Transport and De-Glycosylation

In order to become accessible to cytosolic proteasomes, integral proteins have to be extracted from the ER membrane prior to degradation. Additionally, there should exist a machinery, which shuttles luminal substrate proteins into the cytosol. This transport apparatus must be capable to confer selective and vectorial movement of proteins through a lipid bilayer, which is opposite to protein import into the ER. The first hint on the nature of this export machinery came from the investigation of a membrane bound ER substrate. Wiertz et al. noticed that during breakdown of MHC-class I molecules in cells expressing the HCMV US2 gene-product a de-glycosylated form of the substrate could be co-precipitated with subunits of the Sec61 translocon (Wiertz et al. 1996b). This association was only observed in the presence of proteasome inhibitors indicating that it indeed represented a MHC class I degradation intermediate. The engagement of Sec61 in ER protein degradation was demonstrated in the yeast *Saccharomyces cerevisiae*. Breakdown of CPY* was shown to be diminished in mutants of Sec61 and Sec63, respectively, under conditions where the import into the ER is not affected (Plemper et al. 1997). Mutations in other components of the translocon impaired the import, but had no effect on CPY* degradation. Similar results were obtained using the yeast reconstituted system. After import, pre-pro-α-factor could be chemically crosslinked to mutant versions of Sec61, which were shown to selectively inhibit retrograde transport but not to affect protein import into the ER (Pilon et al. 1997). These findings suggested a transient interaction of ER degradation substrates and the Sec61 translocation pore during retrograde transport. An association of degradation intermediates with Sec61 may indicate that these substrates were still in the process of import into the ER and had in fact never been released from the translocon. A CPY* mutant harboring an additional glycosylation site at the very carboxy-terminus was completely glycosylated before degradation, demonstrating that this molecule had indeed been fully translocated into the ER lumen and dissociated from the Sec61 import machinery before degradation (Plemper et al. 1999a). Recently, an interaction of CFTR ΔF508 with Sec61ß, which was dependent on the presence of proteasome inhibitors, was reported (Bebök et al. 1998). The Sec61-containing export machinery seems therefore to mediate the extraction of membrane bound substrates was well as the dislocation of soluble luminal proteins. Studies on the Sec61 translocon revealed that in vitro the pore is equally traversible for proteins in both ways (Johnson and van Waes, 1999). Therefore, the Sec61 channel is likely to require additional factors, which determine the design as an import channel or as a dislocation machinery, respectively. It was speculated that ER membrane proteins like the previously characterized Hrd1/Der3, Hrd3 or Der1 (see above) interact with Sec61 and thereby modulate the channel's activity for retrograde transport. Recent work has led to the isolation of conditional Sec61 mutants in yeast that specifically affect protein dislocation from the ER into the cytoplasm but do not impair translocation into the ER (Wilkinson et al. 2000; Zhou and Schekman, 1999). These mutations may provide a valuable tool to elucidate the role of Sec61 in retrograde transport and to isolate interacting factors, which contribute to the formation of the protein export channel.

Proteasomal degradation of glycoproteins is preceded by the removal of the glycan-residues. The characterization of a cytosolic N-glycanase activity (Suzuki et al. 1994) and the finding that the Sec61 pore is large enough to mediate transport of glycosylated polypeptide chains (Johnson and van Waes, 1999) has led to the common view that the removal of sugar residues before proteolytic breakdown takes place in the cytoplasm. Indeed, glycosylated and poly-ubiquitinated forms of CPY* have been found at the cytoplasmic side of the ER membrane in yeast (Hiller et al. 1996). Proteolysis of MHC class I molecules after DTT treatment and in the presence of proteasome inhibitors resulted in a glycosylated intermediate associated with Sec61β (Wiertz et al. 1996b). However, there seems to exists no strict order of de-glycosylating and ubiquitinating reactions for a given substrate, because MHC class I molecules in HCMV US2 expressing cells were found to be already de-glycosylated before ubiquitination (Wiertz et al. 1996b, Shamu et al. 1999).

3.6 Ubiquitination

The attachment of poly-ubiquitin to substrate proteins is a prerequisite for efficient recognition and subsequent proteolysis by the 26S proteasome. In a cell line defective in E1 function, CFTR ΔF508 was not efficiently degraded (Ward et al. 1995). Overexpression of the dominant negative ubiquitin mutant UbK48R, which is impaired in the formation of polyubiquitin chains, diminishes the turnover of ER proteins in mammalian (e.g. Jensen et al. 1995; Ward et al. 1995; deVirgilio et al. 1998) and yeast cells (e.g. Biederer et al. 1996; Hiller et al. 1996). Finally, proteolysis of unstable yeast ER proteins is significantly impaired in the absence of the E2 enzymes Ubc6 and Ubc7, respectively (Biederer et al. 1996; Hiller et al. 1996). Surprisingly, deletion of *UBC6*, *UBC7* or the disruption of both genes does not completely abolish the degradation of ER substrates (Biederer et al. 1997). This raises the question, whether other, so far unknown, E2s contribute to poly-ubiquitination of ER proteins. Recently, yeast Vph1p has been shown to be degraded independently from Ubc6p and Ubc7p function (Hill and Cooper, 2000). There may also exist an additional ER degradation mechanism, which does not involve E2 functions. For example, yeast pre-pro-α-factor was shown to be degraded in vitro in an ubiquitination-independent manner, although proteolysis still depended on an active proteasome (McCracken and Brodsky, 1996; Werner et al. 1996).

3.7 Vectorial Transport – A Function for Poly-Ubiquitination in Protein Dislocation?

The translocation of proteins through membranes has been shown to rely on the hydrolysis of nucleotide-tri-phosphates, which provides the driving force for protein transport and ascertains the vectorial nature of this process (Schatz and Dobberstein, 1996). In bacteria the SecA protein pushes substrates through a channel in the cell membrane, thereby hydrolyzing ATP. Co-translational protein import into the ER is accompanied by GTP hydrolysis of the translating ribosome, which pushes the emerging nascent chain through the ER translocon. In the case of post-translational ER import and the translocation into mitochondria the substrate proteins are supposed to traverse the translocation channel with the help of a pulling mechanism excerted by ATP hydrolyzing chaperones on the other side of the membrane.

So far, the nature of the driving force mediating retrograde transport of ER degradation substrates is unknown. An involvement of ER chaperones in a transport machinery analogous to bacterial SecA is unlikely. SecA shares no homology to the ER chaperones described so far, and an enzymatic activity, which parallels the reversible membrane insertion of SecA, has not been characterized from the ER. Furthermore, ER chaperones have an essential function in protein import, which makes an active role in the transport of proteins in the opposite direction unlikely. Cytosolic Hsp70 and Hsp90 have been shown to contribute to the folding of the cytosolic domains of integral ER membrane proteins, but so far there exists no evidence for an involvement of these enzymes in the dislocation of ER proteins. For example, degradation of luminal pre-pro-α-factor occurred independently from cytosolic Ssa1p in vitro, indicating that this chaperone was not required for protein export (Brodsky et al. 1999).

There exists growing evidence that the attachment of polyubiquitin on the cytosolic side of the ER membrane plays a crucial role in the retrograde transport of ER proteins. Deleting the genes for Ubc6, Ubc7 and Cue1, which should affect the ubiquitin conjugating activity at the yeast ER membrane, resulted in the accumulation of CPY* within the ER (Biederer et al. 1997; Bordallo et al. 1998). This result indicates that proper export of ER proteins destined for degradation depends on the conjugation with polyubiquitin. Indeed, similar results have been obtained for the dislocation of membrane bound degradation substrates. The retrograde transport and subsequent proteolysis of a mutant version of ribophorin A and of unassembled TCRα was impeded after overexpression of UbK48R (deVirgilio et al. 1998; Yu and Kopito, 1999). Yet, the role of polyubiquitination in protein dislocation is not clear. Covalent attachment of ubiquitin to a degradation substrate emerging from the translocation pore may assist to anchor this protein at the cytosolic side of the ER membrane and prevent it from slipping back into the ER. Subsequently, such arrested intermediates may be pulled out of the ER by a so far unknown mechanism and degraded. Polyubiquitination of substrate proteins may thus work like a molecular ratchet and assist to define the direction of retrograde transport. Additionally, attachment of ubiquitin molecules to the cytoplasmic domains of integral ER proteins was proposed to initiating dislocation of these substrates from the membrane. In the case of virally induced degradation of MHC-class I molecules this has been shown not to be the case: Mutations replacing lysine residues present in the cytoplasmic domain of MHC-class I molecules did not affect extraction and degradation of this protein (Shamu et al. 1999). However, as mentioned above HCMV induced turnover of MHC-class I molecules exhibits mechanistic differences when compared to the breakdown of ER proteins sorted out by the quality control system. In addition to the described mechanism, there may exist pathways for protein export from the ER, which do not depend on polyubiquitination. Yeast pre-pro-α-factor was shown to be degraded in a proteasome dependent manner in vitro, however export and proteolysis occurred independent from polyubiquitination (McCracken and Brodsky, 1996; Werner et al. 1996).

Jentsch and co-workers proposed a direct involvement of the proteasome in the retrograde transport of ER proteins (Mayer et al. 1998). Breakdown of an artificial substrate protein, which was composed of a non-ER degradation signal fused to a membrane anchor, occurred in two steps: In proteasomal mutants the cytosolic part of the fusion protein was degraded more rapidly than the transmembrane segment which resulted in the accumulation of a proteolytic fragment (Mayer et al. 1998). This experiment indicated that the proteasome is able to extract a protein from the ER membrane and raised the speculation that the ATPases located in the 19S cap particle of the proteasome may be a part of an ER protein dislocation machinery. Such a function would also explain the need for poly-ubiquitination in the dislocation of ER proteins, because the 26S proteasome displays high affinity only for poly-ubiquitinated substrates. However, as Mayer et al. used a non-ER degradation signal to initiate proteolysis of their substrate, the impact of their results on the breakdown of other ER substrates remains to be determined. Indeed, the degradation of other ER membrane proteins (e.g. CFTR, Pdr5*, Sec61-2) does not involve a proteolytic intermediate. Very recently, the retrograde translocation from the ER preceding proteasomal degradation of unassembled immunoglobulin light chains has been shown to depend on the proteolytic activity of the proteasome (Chillaron and Haas, 2000). This again suggests an involvement of this protease in the export process of ER proteins.

3.8 Degradation by the Proteasome

The involvement of the 26S proteasome in the degradation of ER proteins was demonstrated by the use of specific inhibitors in mammalian cells (Jensen et al. 1995; Ward et al. 1995) and the investigation of conditional proteasome mutants in yeast (Biederer et al. 1997; Hiller et al. 1996). It is still an open question, how substrate proteins, that have been dislocated from the ER, are targeted to the degrading 26S proteasome. Several observations suggest a localization of 26S proteasomes at the cytosolic side of the ER membrane. Impairing the catalytic activity of the proteasome by specific inhibitors results in the accumulation of ubiquitinated forms of CFTR ΔF508 in insoluble structures surrounded by ER membranes (Johnston et al. 1998).

A closer investigation of these aggregates revealed, that they also contained cytosolic Hsp70, Hsp90 and at least some proteasomal subunits suggesting that degradation of CFTR ΔF508 is catalyzed by proteasomes localized at or near to the ER (Wigley et al. 1999). Indeed, an association of the 26S proteasome with ER membranes in yeast and permeabilized rat liver cells, respectively, was recently observed (see later; Enenkel et al. 1998; Sakata et al. 1999). It is not clear, whether this association is mediated by the interaction with specific receptors or if proteasomes are recruited by other mechanisms to the ER. For example, the appearance of poly-ubiquitinated substrate proteins associated with the ER may target 26S proteasomes, which bind to polyubiquitin with high affinity, to this membrane. However, the proteolysis of ER substrate proteins need not necessarily involve membrane bound 26S proteasomes. MHC class I molecules, for example, accumulate in HCMV US11 expressing cells and the presence of proteasome inhibitors as soluble ubiquitinated intermediates in the cytosol indicating that this substrate may be degraded by proteasomes, that are not associated with the ER (Wiertz et al. 1996b; Shamu et al. 1999). However, protein breakdown mediated by viral gene-products is likely to follow different pathways than the turnover of misfolded ER degradation substrates and the usage of proteasomal inhibitors may affect the cellular localization of substrate proteins and the 26S proteasome.

It has not been conclusively ruled out that in addition to the 26S proteasome other proteases contribute to the degradation of ER proteins. A function of the signal peptidase, which removes the amino-terminal signal sequences during ER protein import, in the proteolysis of ER membrane proteins has been proposed. Lodish and co-workers observed that the breakdown of an unassembled subunit of the asialoglycoprotein receptor involves a proteolytic intermediate that was most likely generated by the enzymatic action of the signal peptidase complex (SPC) (Yuk and Lodish, 1993). The cleavage site, which led to the formation of the breakdown fragment, shared similarity with the processing sites for SPC in signal sequences. Moreover, proteolytic breakdown of an artificial fusion protein was significantly delayed in mutants of the Sec11 and Spc3 subunits of the signal peptidase, respectively (Mullins et al. 1995; Fang et al. 1997). SPC does not seem to constitute an essential factor for the breakdown of other degradation substrates and may thus contribute only to the turnover of a small subset of ER proteins. A putative protease, termed ER-60, from the ER of mouse liver cells was recently identified, although an involvement of this enzyme in protein degradation has not been convincingly demonstrated (Otsu et al. 1995).

3.9 Cellular Role of ER Protein Degradation

Although ER protein degradation seems not be essential for yeast cells, the breakdown of mutated and thus malfolded ER proteins is often associated with severe diseases in human (Ciechanover, 1998; Plemper and Wolf, 1999). The importance of this process in the understanding of genetic disorders like cystic fibrosis and α1-antitrypsin deficiency has been outlined above. In the following we will give further examples on how viruses or toxins may misuse the machinery for the ER protein degradation to interfere with cellular processes.

3.9.1 Vpu Mediated Degradation of CD4

HIV has developed a method to reduce the amount of its receptor protein CD4 at the plasma membrane of infected cells. The virus encodes a protein Vpu that inserts into the ER membrane and targets CD4 for destruction (Bour et al. 1995; Schubert et al. 1998). Interestingly, the Vpu protein contains a recognition signal for the human-ß-transducin-repeats-containing protein (h-βTrCP), which is also found in cellular proteins like β-catenin and IκBα (Laney and Hochstrasser, 1999). H-βTrCP, a F-box and WD domain containing protein, binds to this signal in a phosphorylation-dependent manner and induces ubiquitination followed by proteolysis of the substrate. In this case, however, the h-βTrCP signal does not initiate breakdown of Vpu, but rather induces the proteolysis of interacting CD4. Expression of Vpu causes the formation of a protein complex that contains CD4, Vpu and the h-βTrCP (Margottin et al. 1998). This complex in turn is thought to associate with SKP1, a component of a cytosolic E3 complex, which mediates ubiquitination and subsequent proteolysis of CD4. Thus, Vpu seems to induce the degradation of CD4 by a mechanism, which normally mediates the breakdown of soluble non-ER proteins. It remains to be determined, whether Vpu dependent CD4 turnover matches the path observed in the proteolysis of misfolded proteins of the ER or represents a new system of ER protein degradation. Breakdown of the CD4 receptor in HIV infected cells is thought to prevent viral super-infection and also to promote the maturation and release of newly assembled virus particles.

3.9.2 Toxins

Several cellular toxins of the AB type have been shown to enter the cell by endocytosis, followed by intracellular transport to the Golgi compartment and the ER (Hazes and Read, 1997). Among this subgroup are the cholera toxin, heat-labile enterotoxin, pertussis toxin, shiga toxin and ricin. Within the ER, these toxins get activated by the formation of disulfide bonds. However, the mode of translocation through the ER membrane into the cytosol, where they exert their toxic function, is unclear so far. Several observations led to the speculation, that toxins may utilize the Sec61 retrograde transport machinery to enter the cytoplasm (Hazes and Read, 1997; Lencer et al. 1999). Export of a heterologously expressed version of the plant toxin ricin subunit A, which was translocated into the ER of yeast cells by fusion to a signal sequence, was significantly reduced in cells expressing proteasomal mutants as well as Sec61 alleles, that were shown to specifically inhibit retrograde translocation of ER degradation substrates (Simpson et al. 1999). Even though, transport was not affected by a deletion of *UBC6* and *UBC7* a large portion of the toxin was degraded by the 26S proteasome after membrane traversal. It was speculated that the remarkable low content of lysine residues in the A subunits may enable a sufficient number of molecules to escape ubiquitination and proteolysis. Recent work revealed an association of ricin subunit A with Sec61α in mammalian cells (Wesche et al. 1999) and a dependence of cholera toxin A1 transport on Sec61 complex function in vitro (Schmitz et al. 2000), which further supports the idea, that AB-type toxins enter the cytosol by a mechanism related to the dislocation of ER protein degradation substrates.

4 A function of Ubiquitin-Conjugation at the Cell Surface

4.1 Signal Transducing Receptors Are Down-Regulated by Internalization

The activity of receptors, transporters or ion channels of the plasma membrane is tightly regulated. Very often, such a regulation comprises variations in the abundance of these proteins at the cell surface. Therefore, controlling the rates of protein internalization is an important tool for regulation of activities in the plasma membrane. In the case of signal transducing receptors, particularly, cells return to an unstimulated stage through accelerated endocytosis of the involved receptors, a mechanism known as down-regulation. The importance of these regulatory issues is underlined by the fact that a misregulation will cause severe diseases. For example, an inherited form of hypertension is caused by the lack of internalization of an epithelial Na+ channel in human kidney cells (Snyder et al. 1995). Reduced internalization of the epidermal growth factor receptor (EGF receptor) results in phenotypes characteristic of transformed cells (Vieira et al. 1996).

Once internalized, receptors can either be transported to the lysosome for degradation or be recycled back to the cell surface.

Endocytosis is initiated by modification of the involved receptor in response to certain stimuli. One of these modifications is the attachment of ubiquitin. The first membrane proteins that were shown to be conjugated with ubiquitin were the platelet-derived growth factor β-receptor (PDGFRβ) and the growth hormone receptor (GHR), respectively (Yarden et al. 1986; Leung et al. 1987). When these receptors were sequenced, a second N-terminal sequence was observed that correlated with that of ubiquitin. The T-cell receptor was among the first examples for which it was shown that ligand binding triggered ubiquitin-conjugation on the cytosolic tail (Cenciarelli et al. 1992). Similar results have been observed for other receptors (see Table 2). The conjugation of PDGFRβ with ubiquitin was dependent on the kinase activity of the receptor (Mori et al. 1993). In the case of the high-affinity receptor for IgE (FcεRI receptor) and the T-cell receptor (TCR)

Table 2. Ubiquitinated plasma membrane proteins

Protein	Reference
In yeast	
ABC peptide transporter Ste6	Kölling and Hollenberg, 1994
Multidrug transporter Pdr5	Egner and Kuchler, 1996
α-factor receptor Ste2	Hicke and Riezman, 1996
a-factor receptor Ste3	Roth and Davis, 1996
Uracil permease Fur4	Galan et al. 1996; Marchal et al. 1998
Galactose permease Gal2	Horak and Wolf, 1997
Maltose transporter	Lucero et al. 2000
General amino acid permease (Gap1)	Springael and André, 1998
In mammalian cells	
PDGF receptor	Yarden et al. 1986
Prolactin receptor	Cahoreau et al. 1994
T-cell receptor	Cenciarelli et al. 1992
Fcε receptor 1	Paolini and Kinet, 1993
SLF receptor (c-kit)	Miyazawa et al. 1994
EGF receptor	Galcheva-Gargova et al. 1995
FGF receptor	Mori et al. 1995
CSF-1 receptor	Mori et al. 1995
Growth hormone receptor	Strous et al. 1996
Rhodopsin	Obin et al. 1996
p185 (c-erbB-2)	Mimnaugh et al. 1996
Met tyrosine kinase receptor	Jeffers et al. 1997
Epithelial Na-channel (ENaC)	Staub et al. 1997
Complement receptor 2	Hein et al. 1998

the modification with ubiquitin was reversible upon disengagement of the ligand (Paolini and Kinet, 1993).

The function of ubiquitin-conjugation during endocytosis remained obscure for many years, although it became clear that, in contrast to ER-degradation, it did not result in degradation by the 26S proteasome (Hicke, 1999; Strous and Govers, 1999). In 1996 experiments performed in yeast pointed to a very unexpected function of ubiquitin-conjugation at the plasma membrane: It constitutes a signal for the endocytosis of cell surface receptors (Hicke and Riezman, 1996).

4.2 Ubiquitin-Conjugation Serves Essential Functions for Internalization of Receptors of the Yeast Plasma Membrane

The first yeast membrane protein, which was shown to be a target of ubiquitin conjugation, was the ABC-transporter Ste6. Kölling and Hollenberg demonstrated that high molecular weight forms of this protein accumulated in mutants affected in endocytosis (Kölling and Hollenberg, 1994). Surprisingly, the short-lived Ste6 was stabilized in cells lacking the ubiquitin-conjugating enzymes Ubc4/Ubc5 but also by mutations in proteases of the vacuole, the lysosome-like compartment of yeast. Later, other proteins of the yeast plasma membrane where found to be ubiquitin-conjugated, too (Table 2). Among them are the G-protein-coupled signal-transducing receptors for the yeast mating factors α (Ste2) and a (Ste3) (Hicke and Riezman, 1996; Roth and Davis, 1996). Interaction of these receptors with their ligands initiates a signaling cascade required for the mating reaction and stimulates the internalization of the receptors. Both receptors are ubiquitin-conjugated constitutively. However, Hicke and Riezman could show that ligand binding stimulates phosphorylation and further ubiquitination of the α-factor receptor. Modified Ste2 is transported to the vacuole where the receptor is permanently inactivated by degradation. The proteasome is not required for this proteolysis. It was concluded, that ubiquitin-conjugation may either promote endocytosis or the vacuolar degradation of Ste2. Since ubiquitination was increased in an end4 mutant, which blocks endocytosis, it was assumed that the modification occurred at the plasma membrane. Evidence pointing to a function of ubiquitination in internalization came from experiments in which ubiquitination was abrogated: Mutants in UBC4/UBC5 are specifically unable to internalize Ste2. Further proof for such a function was provided by the analysis of Ste2 mutants that do not contain cytoplasmic lysine residues. These modified receptors cannot be conjugated with ubiquitin and are not internalized and degraded. Roth and Davis have made similar observations for the a-factor receptor (Roth and Davis, 1996; Roth et al. 1998). Besides these two receptors, a number of transporters and permeases have been shown to undergo ubiquitin-dependent internalization (Hicke, 1999; Strous and Govers, 1999). Moreover, experiments with the general amino acid permease (Gap1) and with the uracil permease (Fur4) have implicated a ubiquitin ligase of the HECT family, Rsp5, in this process (Hein et al. 1995; Galan et al. 1996). These data demonstrated for the first time that ubiquitin-conjugation had a function other than tagging a protein for degradation by the proteasome.

Obviously, these results raise the question why ubiquitin-conjugated receptors are not degraded by the proteasome. Proteasomal degradation of a membrane protein has two requirements: i. It would require the attachment of a Lys48 linked poly-ubiquitin chain (Pickart, 1997) and ii. Most likely it would be dependent on an export channel, as it was demonstrated for ER-bound membrane proteins (see above and Sommer and Wolf, 1997). However, most receptors are modified with only mono- or di-ubiquitin, which was demonstrated with cells lacking the de-ubiquitinating enzyme Doa4. In this mutant, the formation of poly-ubiquitin chains is largely reduced due to a reduction in the pool of free ubiquitin. Ste2 internalization is abrogated in *doa4* cells, but can be restored by expression of additional ubiquitin. When lysine-free versions of ubiquitin, which do not allow the formation of poly-ubiquitin chains, are expressed in the doa4 mutant, endocytosis of Ste2 is restored. In similar experiments, versions of Ste2 have been used that contain only one lysine residue in the cytoplasmic tail. Also in this case, internalization could be restored with the lysine-free ubiquitin (Terrell et al. 1998). These results confirm that mono-ubiquitination on a single lysine residue is sufficient for rapid internalization of Ste2. In support of this, it was shown that also the a-factor receptor is mono- or di-ubiquitinated (Roth and Davis, 1996). Slightly different results have been found for the uracil permease (Fur4), a protein that undergoes regulated endocytosis. This protein was modified with oligo- or poly-ubiquitin chains linked through Lys63. Mono-ubiquitination seems to be sufficient for internalization of Fur4 but formation of the Lys63 linked chain increases the efficiency of endocytosis (Galan and Haguenauer-Tsapis, 1997). Taken together, it can be assumed that the ubiquitin modification of these receptors at the cell surface is different from a modification that leads to proteolysis by the 26S proteasome.

In many cases of regulated protein degradation it was shown that ubiquitination is positively regulated by phosphorylation. This holds true also for endocytosis. Internalization of the α-factor receptor, for example, was shown to depend on specific serine residues in the cytoplasmic tail. They are found in a 9 amino acid motive, SINNDAKSS, which is essential for endocytosis. Mutant versions of Ste2 in which the three serine residues have been eliminated are not phosphorylated, ubiquitin-conjugated and internalized. In agreement with that, mutants in casein kinase I, the kinase which phosphorylates Ste2, are deficient in phosphorylation, ubiquitination and internalization of the receptor (Hicke et al. 1998). Consequently, it can be assumed that phosphorylation precedes ubiquitination. Similarly, internalization of the uracil permease is also triggered by phosphorylation. In this case, a PEST like sequence is required for phosphorylation, ubiquitin-conjugation and internalization (Marchal et al. 1998). Also Ste3 carries such a PEST like sequence which is required for ubiquitin-conjugation and endocytosis. However, it is not required for phosphorylation of the receptor (Roth and Davis, 1996; Roth et al. 1998).

The experiments described above clearly demonstrate a function of ubiquitin-conjugation in endocytosis. However, it remained unclear how ubiquitin triggers internalization. It could either induce a conformational change in the target protein, which exposes a previously masked endocytosis signal, or ubiquitin itself may contain the signal for internalization. Recently, experiments have been published that strongly support the second hypothesis. Truncated versions of the Ste2 receptor lacking cytoplasmic sites for ubiquitin-conjugation are internalized when ubiquitin is fused to them. The SINNDAKSS element is not necessary for endocytosis of Ste2 chimera that contain ubiquitin moieties. If ubiquitin would indeed serve as a signal for endocytosis it should be transferable signal. This was tested with the plasma membrane proton ATPase, encoded by *PMA1*. This protein undergoes rapid internalization when fused to sequences of Ste6 or Ste3 that are needed for ubiquitination. In an alternative approach, Hicke and coworkers fused ubiquitin to Pma1 and observed degradation in the vacuole, which was dependent on endocytosis (Shih et al. 2000). Taken together, these observations indicate that ubiquitin itself carries signals for endocytosis. However, the signal could not be confined to a linear stretch of amino acids. Instead, the three-dimensional structure of ubiquitin seems to be important and especially the residues Ile44 and Phe4 are essential for endocytosis. Furthermore, it can be speculated that phosphorylation of the receptors is only required to trigger ubiquitin-conjugation (Shih et al. 2000).

4.3 Ubiquitin-Dependent Down-Regulation of Receptors in Mammalian Cells

Although ubiquitin-conjugation of receptors in mammalian cells has been observed, the function of this modification seems less clear. In the case of the growth hormone receptor (GH-receptor) and the epithelial Na+ channel (ENaC), it seems likely that ubiquitin-conjugation triggers their endocytosis.

For the GH-receptor, the most thoroughly studied example, Strous and co-workers have investigated ligand stimulated endocytosis of this receptor in a cell line expressing a temperature sensitive ubiquitin-activating enzyme. In these cells, internalization is impaired and the receptor accumulates at the cell surface in a non-ubiquitinated form, suggesting that endocytosis and ubiquitin-conjugation are related (Strous et al. 1996). Surprisingly, it was found that the receptor undergoes ligand stimulated ubiquitin-conjugation, but the lysine residues present in the cytoplasmic tail were not essential for ubiquitin-dependent internalization. Instead, a 10 amino acid motive lacking lysine residues was found to be necessary both for ubiquitin-conjugation and endocytosis (Govers et al. 1999). Moreover, inhibitors of the proteasome blocked both degradation and endocytosis of the wild type as well as the lysine-less variant (van Kerkhof et al. 2000). Thus, it was speculated that ubiquitin-conjugation and/or proteasomal removal of an unknown factor is required prior to endocytosis of the GH-receptor. Its own ubiquitin-conjugation might be a byproduct of the interaction with the conjugation machinery mediated by the 10 amino acid motive.

Another ubiquitin-conjugated protein of the plasma membrane is ENaC. Increased activity of this channel at the cell surface causes an inherited form of hypertension known as Liddle's syndrome. All identified mutations causing this syndrome are concentrated in a proline-rich motive of the cytoplasmic tails of the channel β and γ subunits. In the wild-type channel these motives mediate the interaction with Nedd4, the mammalian homologue of the ubiquitin ligase Rsp5 (Staub et al. 1996). Furthermore, it was demonstrated that ENaC is poly-ubiquitinated and that mutations in the cytoplasmic lysine residues reduce the degradation rates (Shimkets et al. 1997; Staub et al. 1997). Thus, the generation of this human disease seems to be directly linked to a disturbed interaction of the channel protein with the ubiquitin-conjugation machinery.

Protein degradation of ENaC, the GH-receptor, and a number of other proteins including the PDGF-receptor, the Met-receptor and p185c-erbB2 proto-oncogene is delayed upon treatment with inhibitors specific for the proteasome as well as for the lysosomal proteases (Hicke, 1999; Strous and Govers, 1999). This observation can be explained in two ways. First, it is possible that different fractions of these proteins are degraded through different proteolytic systems. Alternatively, the proteasome might attack parts of the proteins that are orientated towards the cytosol while those parts that are located to the lysosomal lumen are directly digested by the resident proteases. This is in contrast to observations with the yeast G-protein-coupled receptors, which seem to be degraded exclusively by vacuolar proteases (Hicke and Riezman, 1996; Roth and Davis, 1996). However, studies with yeast Ste6 protein also pointed to an influence of the proteasome (Loayza and Michaelis, 1998).

Recent experiments have addressed the nature or the mammalian machinery involved in modifying proteins of the plasma membrane with ubiquitin. Yarden and co-workers have established an in vitro reconstituted system to ubiquitinate and degrade the EGF receptor (Levkowitz et al. 1999). It consists of purified ubiquitin-activating enzyme, ubiquitin-conjugating enzyme H5B or H5C and of a known factor, the c-Cbl protein. Previously, it has been shown that overexpression of c-Cbl increased the down-regulation of EGFR and PDGFR. C-Cbl is a protein of 120 kDa that contains an N-terminal SH2 domain and a C-terminal proline-rich region and several tyrosine phosphorylation sites. Between these region a C3HC4 type RING finger is found. Cbl-b and Cbl-3, the two other members of this protein family, also contain the RING finger but in the two oncogenic forms, v-Cbl and 70Z-Cbl it is corrupted. While all three members of the Cbl family are able to mediate ubiquitination of EGFR the oncogenic forms are not. Based on the in vitro observations a model was postulated how ubiquitin-conjugation of EGFR is regulated. Binding of EGF to the extracellular part of the receptor stimulates its kinase activity and results in phosphorylation of the receptor at Tyr-1045. This creates a binding site to which c-Cbl binds via its SH2 domain. Bound c-Cbl becomes phosphorylated at a site near the RING finger domain. Modified c-Cbl is now able to recruit the respective E2 enzyme to the complex, which in turn mediates the conjugation of the receptor with ubiquitin. Since RING finger containing proteins have been found as subunits of E3 complexes, it is feasible to speculate that also c-Cbl might function as part of an E3 complex. Thus, the c-Cbl mediated degradation of EGFR might represent the first example of a regulated ubiquitin-conjugating event in which both the substrate as well as the E3 complex are activated by tyrosine phosphorylation. The c-Cbl docking site in EGFR is found in a similar manner in Erb-B1 and Erb-B2. Both proteins undergo ubiquitination and lysosomal degradation. However, it is absent from Erb-B3, a receptor that is not modified with ubiquitin and constitutively shuttles between endosome and plasma membrane. Most interestingly, EGFR mutants that are unable to interact with c-Cbl and thus exhibit no ubiquitination or degradation, constantly shuttle between cell surface and an early endosomal compartment. In consequence, Yarden and co-workers speculated that ubiquitination plays a critical role in sorting of EGFR from the early endosomal compartment to the lysosomes rather than during endocytosis at the plasma membrane (Yarden et al. 1999).

5 Subcellular Distribution of Components of the Ubiquitin-Proteasome System

There has been rapid progress in understanding the involvement of the ubiquitin-proteasome system in the metabolism of eukaryotic cells, but some important questions on the cellular localization of the degradation machinery still remain. The general view for many years was that the activity of the ubiquitin-proteasome pathway is limited to the cytosol and to the nucleus. Despite lacking direct evidence for nuclear degradation pathways, this idea was supported by the occurrence of poly-ubiquitinated proteins in both compartments. However, some recent reports have demonstrated specific nuclear pathways and pointed to a linkage of ubiquitin-mediated protein degradation to nuclear transport functions. In this context, two complexes of questions are of special interest: First, in which cellular compartment do poly-ubiquitination and final breakdown of the target proteins actually occur? Is for a given nuclear protein the site of its function identical with the site of its ubiquitination and/or degradation? And, second: What are the relevant nucleo-cytoplasmic transport pathways and how are they linked to the ubiquitin-proteasome pathway to gain spatial control on the degradation machinery?

5.1 Ubiquitin-Activating Enzyme, Ubiquitin-Conjugating Enzymes and Ubiquitin-Ligases

In mammalian cells the E1 ubiquitin-activating enzyme exists in two isoforms, E1a (110 kDa) and E1b (117 kDa), which are derived from a single gene and mRNA (Cook and Chock 1992, Handley Gearhart et al. 1994). The isoform E1a is predominantly found in the nucleus and has been shown to harbor a functional nuclear localization sequence (NLS) required for nuclear targeting and phosphorylation. In contrast, E1b lacks the NLS, is not phosphorylated and localized in the cytoplasm (Handley-Gearhart et al. 1994; Stephen et al. 1997). Phosphorylation of E1a was demonstrated to occur in a cell cycle-dependent manner, being maximal in G2 phase (Stephen et al. 1996). Since the enzymatic activity of E1a was independent from increased phosphorylation the function of E1 phosphorylation remains unclear. From the three E1 homologs in the yeast *Saccharomyces cerevisiae* only Uba2p harbors a putative NLS. Uba2p is largely localized to the nucleus (Dohmen et al. 1995) and was shown to be essential for Smt3p-activation (Johnson et al. 1997).

Although E2s and E3s appear to be generally present in the nucleus and in the cytoplasm (Schwartz and Ciechanover, 1999), this finding does not exclude that the activity of individual E2 and E3 enzymes may be restricted to a certain site. With the exception of the membrane-bound E2s (see below), there is presently no experimental evidence for a functional targeting sequence within a single E2 or E3 that could direct the protein to a distinct subcellular compartment. However, in yeast, immunochemical localization of the gene product and biochemical data have shown that the E2 Ubc3/Cdc34 is a predominantly nuclear protein (Goebl et al. 1994). This finding matches the observation that relevant substrates of Ubc3/Cdc34 reside within the nucleus. Several cell cycle events have been shown to be under control of Ubc3/Cdc34 (Deshaies et al. 1995; Plon et al. 1993; Goebl et al. 1988). Furthermore this E2 regulates the function of the transcripition factors Gcn4p and Rgt1p (Kornitzer et al 1994; Li and Johnston, 1997). In mammalian cells, transfected human Cdc34 has been proposed to serve a function analogous to its yeast counterpart (Pagano et al. 1995; Plon et al. 1993). Human Cdc34 represents a nuclear protein and is suggested to mediate the ubiquitination of the transcripton factor MyoD (Lisztwan et al. 1998; Song et al. 1998). Yeast Ubc4 and Ubc5 are required for the degradation of many abnormal and short-lived proteins including nuclear mitotic cyclins and the Matα2 repressor protein (Hochstrasser et al. 1999). Thus, at least in the latter cases both enzymes should have enzymatic activity on target proteins in the nuclear compartment.

In yeast cells, the E2 Ubc6 is membrane-bound at the ER (Sommer and Jentsch, 1993) and is assumed to be involved in ubiquitination processes at the cytosolic face of the ER. The soluble ubiquitin-conjugating enzyme Ubc7 has to be recruited to the ER surface by its receptor, Cue1 (Biederer et al. 1997), to catalyze ubiquitination and finally proteolysis of both ER proteins and soluble non-ER proteins (see chapter ER degradation). In addition, Ubc7p was also proposed to interact physically with Ubc6p forming a membrane-associated heteromeric complex (Chen et al. 1993). Ubc8 is located in the cytoplasm and one of its function has been shown to reside in catabolite degradation of the gluconeogenic enzyme fructose-1.6-bixphosphatase (Schüle et al. 2000). Ubc9 conjugates the ubiquitin-like proteins SUMO-1/Smt3 (Johnson and Blobel, 1997; Schwarz et al. 1998). From current data it seems to be predominantly nuclear.

Some ubiquitin ligase activities are coupled with large-multisubunit protein complexes. The substrates of these E3 complexes represent a broad spectrum of proteins that participate in a variety of cellular functions, e.g. regulation of CDK activity, activation of transcription, signal transduction, assembly of kinetochores, and DNA replication. Concerning those substrates relevant for the cell cycle the critical issue is how and when these proteins are ubiquitinated. Studies of cell cycle regulation have demonstrated that two E3 complexes play a crucial role for the timing of cell cycle regulator proteolysis: the cyclosome/anaphase-promoting complex (APC) and the Skp1-cullin-F-box protein ligase complex (SCF) (reviewed in Deshaies, 1999; Koepp et al. 1999, Winston et al. 1999). Regulatory proteins of the cell cycle fulfill their function mainly in the nucleus, but at present experimental evidence for specific nuclear pathways of ubiquitin-mediated degradation is lacking. However, in *Saccharomyces cerevisiae*, the feedback-regulated degradation of the transcriptional activator Met4 has been shown to be triggered by the SCF^{Met30} ubiquitin-protein ligase. The F-box protein Met30p itself is short-lived and localizes to the nucleus (Rouillon et al. 2000), also indicating a nuclear localization of the whole complex. From this example we can assume a defined cellular distribution of the E3 complexes and, consequently, a spatial regulation of their activity.

5.2 Proteasomes

As discussed in the previous chapter there is some evidence both for a defined spatial arrangement of the ubiquitin-conjugating machinery and for the compartment-specific activity of several components. For understanding of ubiquitin-proteasome-mediated proteolysis a further fundamental issue is the spatial relationship within the cell between the protease itself and its substrates. Thus the questions arise whether active proteasomes are distributed homogeneously throughout the cell or if there exist defined cellular sites of proteasome-dependent degradation? In general, studies of proteasomal distribution revealed differences in the localization of proteasome subpopulations, demonstrated the functional importance of endoplasmic reticulum-associated proteasomes, and investigated the role of the putative nuclear localization signals on proteasome transport into the nucleus (reviewed in Rivett, 1998). Some still existing disagreement on the details could be explained by a variation with cell types, cell cycle stage and development. For example, a cell-cycle-dependent change in the distribution of nuclear proteasomes was observed in higher eukaryotes, with a co-localization of the proteasome with chromosomes and the spindle during metaphase (Kawahara and Yokosawa, 1992).

In mammalian cells, previous reports proved the localization of proteasomes both in the cytoplasm and in the nucleus (Palmer et al. 1996; Peters et al. 1994). In a more recent study GFP-tagged 20 S proteasomes in fibrosarcoma cells were shown to be dispersed over the cytoplasm and nucleoplasm wherein they seem to diffuse rapidly (Reits et al. 1997). Reits et al. used a fusion protein of GFP with the proteasome subunit LMP2, which replaces a β subunit of the proteasome upon induction with interferon γ, a stimulator of

MHC class I presentation (Belich et al. 1994; Nandi et al. 1996). The authors also demonstrated that proteasomes were transported slowly and unidirectionally from the cytoplasm to the nucleus, but can enter the nucleus rapidly upon its re-assembly during cell division. Several other studies also provided evidence for a preferential localization to nuclear substructures (Grossi de Sa et al. 1988, Pal et al. 1988). Putative nuclear localization signals (NLS) that have been found in some α subunits of the 20S proteasomal core complex (Tanaka et al. 1990), might function in targeting proteasomes into the nucleus, which was shown for single proteasome subunits in vitro (Knuehl et al. 1996; Wang et al. 1997). A distinct fraction of proteasomes was also reported to be associated with the cytoplasmic face of the ER membrane (Rivett et al. 1997; Rivett et al. 1992; Yang et al. 1995). There it is thought to degrade mutant proteins extracted from the ER (see chapter ER degradation) or generate antigenic peptides for presentation on MHC class I molecules (Coux et al. 1996; Lehner and Cresswell, 1996; Schild and Rammensee, 2000; Schubert et al. 2000; Reits et al. 2000).

Recently, nuclear and perinuclear substructures were suggested as sites of concentration of proteasomes. Johnston et al. (1998) reported the formation of a novel pericentriolar structure, termed "aggresome". In transfected human embryonic kidney (HEK293) or Chinese hamster ovary cells overexpression of the integral membrane protein CFTR, which is inefficiently folded and exported from the ER, or inhibition of proteasome activity resulted in the accumulation of stable, ubiquitinated aggregates of CFTR. Once formed, these aggregates are delivered to an ubiquitin-rich structure at the centrosome/microtubule-organizing center (MTOC). The microtubule depolymerizing drug nocodazole prevents the perinuclear formation of aggresomes and causes the formation of smaller protein inclusions dispersed in the periphery of the cells, indicating that protein aggregates move on microtubles to the MTOC to form the aggresome (Garcia Mata et al. 1999; Johnston et al. 1998). Further investigations on the generality and dynamics of aggresome formation suggested that aggresomes can be formed by soluble, non-ubiquitinated proteins as well as by integral transmembrane ubiquitinated ones. In conclusion, aggresomes are proposed as a general response of the cell to the presence of misfolded and aggregated proteins (Garcia Mata et al. 1999). Regarding the formation of protein aggregates within the cell it is noteworthy that neuronal intranuclear inclusions (NIs) were described as a ultrastructural feature of several neurodegenerative polyglutamine repeat expansion disorders, including Huntington's disease and the spinocerebellar ataxias (SCAs). NIs show high ubiquitin content and colocalization with proteasomes. Recent work showed that these inclusions are not required for expanded polyglutamine pathology (reviewed in Floyd and Hamilton 1999).

Cummings et al. (1999) observed that mutant ataxin-1, the expanded protein causing SCA type 1, is selectively resistant to degradation though it is equally well poly-ubiquitinated as the normal protein. Inhibiting proteasome activity promotes mutant ataxin-1 aggregation in transfected cells indicating that ataxin-1 is degraded by the ubiquitin-proteasome pathway.

To study proteasomal degradation of nuclear proteins and viral antigens in human osteosarcoma cells Anton et al. (1999) used a mutated form of influenza virus nucleoprotein (dNP) as a model protein with a nuclear localization sequence that is ubiquitinated and degraded by proteasomes. Immunoflourescence and biochemical analysis revealed that in the presence of proteasome inhibitors dNP accumulates in highly insoluble ubiquitinated and non-ubiquitinated species in nuclear substructures known as promyelocytic leukemia oncogenic domains (POD) and at the centrosome/MTOC. The authors could show by immunoflourescence that dNP recruits proteasomes and the chaperone HSC70 to both sites. Moreover, after restoring proteasome activity while blocking protein synthesis dNP disappeared from PODs and the MTOC resulting in the generation of MHC class I-bound peptide derived from dNP. These findings suggest PODs and MTOC as sites of proteasomal degradation of misfolded NP and probably other cellular proteins and indicate that antigenic peptides are generated at these sites.

By immunoflourescence and cell fractionation Wigley and co-workers (Wigley et al. 1999; Fabunmi et al. 2000) have shown that active proteasomal complexes degrading ubiquitinated protein and proteasome-specific peptide substrates associate with centrosomes in HEK 293 and HeLa cells. The structure is perinuclear, surrounded by endoplasmic reticulum, adjacent to the Golgi and co-localizes with the centrosomal marker γ-tubulin. Besides misfolded CFTR (Johnston et al. 1998; Wigley et al. 1999), by immunocytochemistry a number of other proteasome substrates have been localized to the centrosome, including the tumor suppressor p53 (Brown et al. 1994), cyclins (Koepp et al. 1999), presenilin 1 (Johnston et al. 1998), and IκB (Crepieux et al. 1997). Moreover, the substrates cyclins A, B1, and E are associated with the centrosome in a cell cycle dependent manner (Bailly et al. 1992, Lacey et al. 1999, Hinchcliffe et al. 1999). Thus, the centrosomal co-localization of active proteasomes with multiple physiological substrates not only suggested the centrosome/MTOC as a novel site for proteasome function, but also provided evidence that proteasome-catalyzed proteolysis may be regulated by its selective partitioning to this structure.

The cellular distribution of 26S proteasome subunits in yeast appears to be similar as in higher eukaryotes but there also exist some differences in the details. Indirect immunoflourescence studies localized the GFP-tagged yeast 26S proteasome primarily to the yeast nucleus and the nuclear periphery throughout the cell cycle (Enenkel et al. 1999; Enenkel et al. 1998; McDonald and Byers, 1997; Russell et al. 1999; Wilkinson et al. 1998). Co-immunolocalization studies revealed overlapping localizations of protea-somes and marker proteins of either ER translocon or nuclear pore compo-nents (Enenkel et al. 1998; Wilkinson et al. 1998). In living cells of *Schizosac-charomyces pombe*, GFP-tagged 26S proteasomes were found predominantly at the nuclear periphery both in interphase and throughout mitosis. How-ever, a dramatic change in the localization was observed during meiosis (Wilkinson et al. 1998). Surprisingly, in fission yeast by electronmicroscopy the proteasome was shown to be concentrated on the inner side of the nu-clear membrane (Wilkinson et al. 1998). Together with biochemical frac-tionation experiments the localization data gave evidence that the majority of yeast proteasomes (about 80%) is structurally bound to the contiguous network of nuclear envelope (NE) and ER membranes (Enenkel et al. 1999). These results were also supported by immunoflourescence data provided by (Russell et al. 1999) who estimated that the fraction of cytoplasmic protea-somes represents at most 20% of the total proteasome population. These findings raise not only questions relating to the function of proteasomes at the NE and their targeting to this compartment but also about the cyto-plasmic activity of proteasomes. It is possible that the small cytoplasmic fraction of proteasomes is sufficient to degrade cytoplasmic substrates. This is consistent with the observation that yeast cells continue to grow despite inhibiting of 70–80% of the proteasome activity (Lee and Goldberg, 1997), suggesting that proteasomes should not be rate limiting for degradation. Otherwise multi-ubiquitinated proteins in the cytoplasm are likely to be transported to the NE/rough ER or the nucleus for degradation by the pro-teasome. Additionally, assuming that most short-lived substrates of the proteasome in yeast are cellular regulators functioning primarily in the nu-cleus, these proteins have to be directed to the NE/rough ER before degra-dation. This not only implies that the ubiquitin-conjugation machinery has to be active in the nucleus but also that nucleo-cytoplasmic trafficking is needed for transport of the nuclear substrates to the site of degradation.

In summary, proteasomes are generally found at subcellular sites to-gether with the vast majority of substrates. The intracellular distribution of proteasomes has been characterized as flexible but the sites of proteasomal concentration are strictly associated with proteolytic activity. This might be due to a guiding process for all degradation components resulting in the formation of large units which could be necessary to facilitate the controlling of the whole proteolytic machinery. Indeed, there is growing evidence for a defined spatial regulation of proteasome-catalyzed proteolysis in yeast and in higher eukaryotic cells. Thus, subcellular distribution, a common mechanism for the regulation of enzymatic systems, appears to be an important feature of regulation of proteasomal proteolysis. However, up to now it is as yet poorly understood what determines the recruitment of proteasomes to different cellular compartments like the NE/rough ER, and if the translocation process represents an active mechanism or simply occurs due to diffusion. The latter possibility is probably unlikely because molecules of the size of the 26S proteasome are thought to be too large to simply diffuse through the cytosol (Janson et al. 1996).

5.3 A Link Between Nucleo-Cytoplasmic Transport and Degradation

A fundamental problem in understanding ubiquitin/proteasome-mediated degradation of regulatory proteins is how individual substrates are selectively targeted at a given time. In principle, protein degradation can be timed by regulating the interaction between the target and components of the ubiquitin system that recognize specific degradation signals or other modifications. The phosphorylation status of the target is suggested to play an important role in selective substrate recognition. Indeed, proteins whose activity is required only at certain stages of the cell cycle are often phosphorylated, modified by covalent attachment of ubiquitin molecules and finally digested by the proteasome (Koepp et al. 1999; Hershko and Ciechanover, 1998). However, some recent reports indicated that phosphorylation followed by polyubiquitination is not sufficient to degrade cell cycle regulatory proteins. Additionally, regulated nuclear import or export has been shown to be involved in the timing of substrate turnover. Considering the preferential localization of proteasomes to the nuclear periphery (see chapter above) the connection between nuclear transport functions and proteasomal degradation is not surprising. With respect to a probable nucleus-specific ubiquitin-proteasome pathway the role of nuclear transport processes in the targeting of substrates and in the selective partitioning of proteasomes is of special interest.

Regulated transport of proteins between the nucleus and the cytosol is determined by their ability to interact with specific nuclear import and export factors. Nuclear import and export proceed through nuclear pore complexes (NPC) (Pante and Aebi, 1996; Doye and Hurt, 1997; Fabre and Hurt, 1997), and can occur via a great number of distinct pathways (reviewed in Görlich and Kutay, 1999; Nakielny and Dreyfuss, 1999). The selective transport of proteins requires energy and depends on targeting signals within the amino acid sequence. Signal-mediated nuclear import is a three-step process: docking at the NPC, translocation, and nuclear deposition of the cargo. In the docking step the cargo in the cytoplasm binds to a soluble import receptor (importin). The cargo-importin complex is targeted to the NPC and translocated to the other side. Upon entering the nucleus the cargo is released and the importin is returned to the cytoplasm for another round of transport. Protein export shares mechanistic similarities with protein import. In the nucleus export receptors (exportins) are able to recognize and bind cargos bearing a nuclear export signal (NES). The complex is then rapidly translocated to the cytoplasm, where it dissociates.

By regulating the transport of proteins into and out of the nucleus eukaryotic cells manage numerous biological processes. Several modes of such regulated nuclear transport had been shown to control cell cycle progression and signalling pathways (reviewed in Kaffman and O'Shea, 1999; Hood and Silver, 1999; Görner et al. 1999; Hopper, 1999). The phosphorylation of the cargo represents a common mechanism regulating the ability of the cargo to interact with specific import or export receptors. Examples for this model of regulation are the yeast transcription factor Pho4 (Kaffman et al. 1998), and the mitotic cyclin B (Hagting et al. 1998, Hagting et al. 1999; Toyoshima et al. 1998; Yang et al. 1998). Cargo-receptor complex formation could also be regulated by intermolecular association of the cargo with itself, with RNA or with other proteins. Other mechanisms of regulated localization are cytoplasmic anchoring, the regulation of the soluble transport machinery and the regulation of the NPC (reviewed in Kaffman and O'Shea, 1999).

Recent studies on cell cycle regulatory proteins have indicated a spatial control of short-lived regulators. For example, Yap1p (Yan et al. 1998), Far1p (Blondel et al. 1999), Cdc6p (Petersen et al. 1999) and Cdc25p (Lopez-Girona et al. 1999) were shown to be down-regulated by specific export from the nucleus. Other reports not only underline the importance of regulated nuclear import and export for controlling the compartment-specific level of regulatory proteins, but also point to the fact that the degradation kinetics of a regulator might be different in the cytosol and the nucleus. First clue for a connection between regulated nucleo-cytoplasmic transport and ubiquitin-mediated degradation came from Loeb et al. (1995) who reported on nuclear transport mutants that could not degrade mitotic cyclins. They showed that the function of Srp1, a yeast importin-α ortholog, is required for the execution of mitosis and suggested that the import of cell cycle regulators into the nucleus is critical for cell cycle progression. In mammalian cells, the down-regulation of two members of the family of basic helix-loop-helix/Per-ARNT-Sim proteins was demonstrated to be ubiquitin-proteasome mediated and to depend on nuclear import and export functions (Davarinos and Pollenz, 1999; Roberts and Whitelaw, 1999). Following exposure to ligands both the aryl hydrocarbon receptor (AHR) and the dioxin receptor (DR) were imported into the nucleus triggering them for phosphorylation and subsequent rapid degradation via the ubiquitin-proteasome pathway. Inhibition of proteasomal function led to nuclear accumulation of both proteins. Stabilization of a constitutively nuclear short-lived form of the DR (DRNLS) was observed when it was co-expressed with the ubiquitin-mutant UbK48R (Roberts and Whitelaw, 1999). In addition, the mutation of a putative CRM1-specific NES present in the AHR or inhibition of CRM1-mediated export by Leptomycin B (Davarinos and Pollenz, 1999) resulted in reduced degradation level and nuclear accumulation. Thus, Davarinos and Pollenz (1999) concluded that ligand bound AHR is degraded in the cytoplasm via the proteasome after being exported from the nuclear compartment. Several other reports have also linked nuclear export with regulated proteolysis. Diehl et al. (1998) reported that ubiquitin-dependent degradation of the cell-regulatory protein cyclin D correlates with its nuclear export. Cyclin D1 accumulates in the nucleus during G1 phase, but it is redistributed into the cytoplasm as cells move through S phase. The authors demonstrated that the specific phosphorylation of cyclin D1 by the glycogen synthase kinase-3β (GSK-3β) not only triggers rapid cyclin D1 turnover by the ubiquitin-proteasome pathway, but also enforces the cytoplasmic localization of cyclin D1. Cyclin D1 does not contain nuclear import or export signals. Thus, Diehl et al. (1998) speculate that GSK-3β-mediated phosphorylation of nuclear cyclin D1 may result in its export by facilitating the interaction with an exportin. Furthermore, the results suggest that GSK-3β actively shuttles between nucleus and cytoplasm and that its observed cell cycle dependent partitioning to the nuclear compartment may be necessary for the relocalization of cyclin D1.

The cyclin-dependent kinase inhibitor p27[Kip1] and the p53 tumor suppressor were also shown to interact with nuclear export mediating factors, which appears to be a requirement for these proteins to be degraded (reviewed in Lain 1999b). Phosphorylated p27[Kip1] is recognized by an SCF-type E3 ligase, which promotes ubiquitination and degradation (Amati and Vlach 1999; Carrano et al. 1999; Sutterlüty et al. 1999; Tsvetkov et al. 1999). Moreover, Tomoda and co-workers (Tomoda et al. 1999) found that phosphorylated p27[Kip1] binds to a novel protein, Jab1, which directs movement of p27[Kip1] from the nucleus to the cytoplasm. This Jab1-dependent relocalization is necessary for proteolysis because a mutant p27[Kip1], which is unable to bind to Jab1, is neither exported nor degraded. This was also supported by the observation that nucleo-cytoplasmic transportation per se is not sufficient for degradation of p27[Kip1], as the fusion of p27[Kip1] with a heterologous NES led to efficient export to the cytosol but not to accelerated degradation. Impairing proteasomal function by specific inhibitors prevented not only the breakdown of p27[Kip1] but also its movement from the nucleus to the cytoplasm. An explanation could be that the proteasome might degrade either a factor interfering specifically with the export of $p27^{Kip1}$ or a more general inhibitor of export. The exact function of Jab1 in $p27^{Kip1}$ degradation remains to be elucidated. It may simply serve as a shuttle for $p27^{Kip1}$, or probably Jab1 itself physically interacts with components of the ubiquitin/proteasome system (reviewed in Scheffner, 1999). The latter possibility seems quite attractive, especially when keeping in mind that Jab1 shows sequence similarity to the yeast protein Rpn11p/Mpr1p, which represents a subunit of the 19S regulatory particle of the proteasome (Lenk et al. unpublished)

The proto-oncoprotein Mdm2 (Freedman et al. 1999) was recently shown to harbor an E3 ubiquitin ligase activity, that directly transfers ubiquitin onto the p53 tumor suppressor protein (Honda et al. 1997). Export of Mdm2 from the nucleus or at least the interaction of Mdm2 with the export machinery is required for the ubiquitin/proteasome dependent degradation of the p53 (Lain et al. 1999a; Freedman and Levine, 1998; Roth et al. 1998; Stommel et al. 1999). Moreover, not only nuclear export but also import of Mdm2 is needed to promote p53 export and degradation, because a mutant Mdm2 protein deficient for nuclear import was unable to reduce p53 levels in transfected cells (Tao and Levine, 1999). Blockage of nuclear export results in accumulation of p53-Mdm2 complexes in subnuclear domains adjacent to the PODs (Lain et al. 1999a), which have been shown to function as sites of proteasomal activity (see above).

Going into some details of these studies there are arising some questions concerning the spatial organization of the ubiquitin-proteasome machinery. The connection of nuclear export and ubiquitin-dependent degradation indicates that proteolysis of the reported substrates itself is a strictly cytoplasmic event. This might be due to the cytoplasmic localization of active proteasomes, of components of the ubiquitin-conjugation system or of factors sensitizing the substrate protein for ubiquitination and/or degradation. However, in response to treatment with specific proteasome inhibitors both p53 and $p27^{Kip1}$ accumulate primarily in the nucleus instead of in the cytoplasm (Freedman and Levine, 1998; Lain et al. 1999a; Smart et al. 1999; Tomoda et al. 1999). These observations can be explained by the assumption, that the exported protein, which is not degraded due to inhibition of cytoplasmic proteasomes, may be rapidly re-imported and therefore accumulate preferentially in the nucleus. However, since proteasomes seem to be present both in the cytoplasm and in nuclear compartments, why degradation of substrates should only occur in the cytoplasm? Scheffner (1999) proposed for $p27^{Kip1}$-breakdown that the export of the substrate is controlled in a proteasome-dependent manner and that ubiquitination and degradation take place in the cytoplasm. However, up to now there exists no experimental evidence for a regulation of nuclear export by the proteasome. Alternatively, breakdown of substrate proteins may not proceed equally distributed over the cytoplasm but rather within the subnuclear or nucleus-associated structures where active proteasomes have been shown to accumulate (see above). In this case, the nuclear substrates must be targeted by nuclear export factors, which facilitate substrate translocation to these proteolytic centers.

Regarding the localization of the majority of proteasomes to the nuclear rim and nuclear envelope, it seems likely, that proteasomal degradation requires nuclear export or translocation processes, which directs nuclear proteins at least to the nuclear pore or to the inner surface of the nucleus. This implies that all components of the ubiquitination machinery have to be active inside the nucleus, which seems to be a prerequisite for the specific export process. It is also possible that nucleus-specific kinases or E2/E3 enzymes promote the triggering of nuclear proteins for rapid degradation.

In summary, the principle, that nuclear proteins might have to be exported for ubiquitin-dependent degradation, seems to act as a common mechanism for controlling the level of short-lived proteins (reviewed in Pines, 1999; Scheffner, 1999; Lain et al. 1999b). With respect to our knowledge so far about the cellular localization of proteasomes and parts of the ubiquitin-conjugation system there remain some still open questions concerning the mechanistic link of nuclear export and degradation via the ubiquitin-proteasome pathway.

5.4 Ubiquitin-Like Proteins

In contrast to ubiquitin, the covalent attachment of ubiquitin-like molecules (Ubls) to other proteins seems to be more important for post-translational protein modification than for protein degradation. There is not much known yet about the subcellular distribution of these proteins but since they are frequently involved in targeting functions it can be assumed that they are also subjected to a spatial organization.

It is known from the interferon-inducible ubiquitin cross-reactive protein (UCRP) that it is conjugated to a small subset of intracellular proteins probably targeting them to the cytoskeleton (Loeb and Haas, 1994).

The function of SUMO conjugation is only beginning to be understood. So far, available data suggest that SUMO-1 is likely to play a role in either targeting modified proteins and/or inhibiting their degradation. This is also supported by the observation that yeast cells deficient in SUMO/Smt3 conjugation are inviable and arrest in the cell cycle before anaphase (Seufert et al. 1995; Li and Hochstrasser, 1999). SUMO-1 conjugation covalently modifies a number of target proteins, including the inflammatory response regulatory protein IkB_ (Desterro et al. 1998), the acute promyelocytic leukemia protein PML (Sternsdorf et al. 1997; Kamitani et al. 1998; Muller et al. 1998), the p53 tumor suppressor (Gostissa et al. 1999), and the GTPase activating protein RanGAP1 (Mahajan et al. 1997; Matunis et al. 1996). SUMO-modified RanGAP1 is recruited to the cytoplasmic face of the nuclear pore where it tightly associates with the nucleoporin RanBP2 (Majahan et al. 1997; Saitoh et al. 1997; Mahajan et al. 1998; Matunis et al. 1998). Thus, the attachment of SUMO-1 might have a targeting function for RanGAP1, which is important for the function of RanGAP1 in nuclear transport. Ran GAP1, together with the Ran GDP/GTP exchange factor, RanGEF, regulates the GTPase Ran, which switches between a GDP- and a GTP-bound form and is required for both nuclear import and export. RanGAP1 and RanGEF are localized to the opposite sides of the nuclear envelope. The asymmetric distribution of both enzymes results in a RanGTP gradient across the nuclear envelope with a high RanGTP concentration in the nucleus and low levels in the cytoplasm (Görlich et al. 1996; Izaurralde et al. 1997). The Ran system likely defines the direction of transport, while the specificity of nuclear import and export is achieved by interaction of signal sequences with importins and exportins. Modification of RanGAP1 by SUMO-1 requires the Ubc9 homologue in Xenopus eggs (Saitoh et al. 1997) and in human cells (Lee et al. 1998). Accordingly, the human homologue of Ubc9 has been shown to co-localize with RanGAP at the nuclear envelope (Lee et al. 1998). Besides its perinuclear localization SUMO-1 also has been shown to be nuclear-diffuse as well as nuclear-punctate, with SUMO-1 concentrated foci corresponding to PML nuclear bodies (Boddy et al. 1996; Duprez et al. 1999; Matunis et al. 1996). Interestingly, in mammalian cells, accumulation of proteasomal substrates together with active proteasomes was observed adjacent to PML oncogenic domains (POD) (Anton et al. 1999; Lain et al. 1999a). Since both the SUMO/Smt3 activating E1 enzyme, Uba2, and the SUMO/Smt3 conjugating E2 enzyme, Ubc9, were localized to the nucleus (see above), covalent modification by SUMO/Smt3 seems to represent a specific modification for nuclear targets. However, in the yeast S.cerevisiae, during distinct phases of mitosis the SUMO homologue Smt3p also has been found specifically attached to septins, which are components of filaments encircling the yeast bud neck (Johnson and Blobel, 1999).

Acknowledgement

The authors would like to thank Drs. Dieter H. Wolf and Linda Hicke for communicating results prior to publication. Dr. Dieter H. Wolf and the members of the Sommer lab are acknowledged for critical comments on the manuscript. This work was partially supported by the Deutsche Forschungsgemeinschaft, the Deutsch-Israelische Projektkooperation (DIP), the Austrian Fonds zur Förderung der wissenschaftlichen Forschung and the Marie Curie TMR program of the European Community.

References

Amati B, Vlach J (1999) Kip1 meets SKP2: new links in cell-cycle control. Nat Cell Biol 1:E91-93

Amerik A Yu, Swaminathan S, Krantz BA, Wilkinson KD, Hochstrasser M (1997) In vivo disassembly of free polyubiquitin chains by yeast Ubp14 modulates rates of protein degradation by the proteasome. EMBO J 16:4826-4838

Anton LC, Schubert U, Bacik I, Princiotta MF, Wearsch PA, Gibbs J, Day PM, Realini C, Rechsteiner MC, Bennink JR, Yewdell JW (1999) Intracellular localization of proteasomal degradation of a viral antigen. J Cell Biol 146:113-124

Ayalon Soffer M, Shenkman M, Lederkremer GZ (1999) Differential role of mannose and glucose trimming in the ER degradation of asialoglycoprotein receptor subunits. J Cell Sci 112:3309-3318

Bailly E, Pines J, Hunter T, Bornens M (1992) Cytoplasmic accumulation of cyclin B1 in human cells: association with a detergent-resistant compartment and with the centrosome. J Cell Sci 101:529-545

Bartel B, Wunning I, Varshavsky A (1990) The recognition component of the N-end rule pathway. EMBO J 9:3179-3189

Baumeister W, Walz J, Zuhl F, Seemüller E (1998) The proteasome: paradigm of a self-compartmentalizing protease. Cell 92:367-380

Bebök Z, Mazzochi C, King SA, Hong JS, Sorscher EJ (1998) The mechanism underlying cystic fibrosis transmembrane conductance regulator transport from the endoplasmic reticulum to the proteasome includes Sec61β and a cytosolic, deglycosylated intermediary. J Biol Chem 273:29873-29878

Belich MP, Glynne RJ, Senger G, Sheer D, Trowsdale J (1994) Proteasome components with reciprocal expression to that of the MHC-encoded LMP proteins. Curr Biol 4:769-776

Biederer T, Volkwein C, Sommer T (1996) Degradation of subunits of the Sec61p complex, an integral component of the ER membrane, by the ubiquitin-proteasome pathway. EMBO J 15:2069-2076

Biederer T, Volkwein C, Sommer T (1997) Role of Cue1p in ubiquitination and degradation at the ER surface. Science 278:1806-1809

Blondel M, Alepuz PM, Huang LS, Shaham S, Ammerer G, Peter M (1999) Nuclear export of Far1p in response to pheromones requires the export receptor Msn5p/Ste21p. Genes Dev 13:2284-2300

Boddy MN, Howe K, Etkin LD, Solomon E, Freemont PS (1996) PIC 1, a novel ubiquitin-like protein which interacts with the PML component of a multiprotein complex that is disrupted in acute promyelocytic leukaemia. Oncogene 13:971-982

Bonifacino JS, Weissman AM (1998) Ubiquitin and the control of protein fate in the secretory and endocytic pathways. Ann Rev Cell Dev Biol 14:19-57

Bordallo J, Wolf DH (1999) A RING-H2 finger motif is essential for the function of Der3/Hrd1 in endoplasmic reticulum associated protein degradation in the yeast Saccharomyces cerevisiae. FEBS Lett 448:244-248

Bordallo J, Plemper RK, Finger A, Wolf DH (1998) Der3p/Hrd1p is required for endoplasmic reticulum-associated degradation of misfolded lumenal and integral membrane proteins. Mol Biol Cell 9:209-222

Bour S, Schubert U, Strebel K (1995) The human immunodeficiency virus type 1 Vpu protein specifically binds to the cytoplasmic domain of CD4: implications for the mechanism of degradation. J Virol 69:1510-1520

Breitschopf K, Bengal E, Ziv T, Admon A, Ciechanover A (1998) A novel site for ubiquitination: the N-terminal residue, and not internal lysines of MyoD, is essential for conjugation and degradation of the protein. EMBO J 17:5964-73

Brodsky JL, McCracken AA (1997) ER-associated and proteasome-mediated protein degradation: how two topologically restricted events came together. Trends in Cell Biol 7:151-156

Brodsky JL, Werner ED, Dubas ME, Goeckeler JL, Kruse KB, McCracken AA (1999) The requirement for molecular chaperones during endoplasmic reticulum-associated protein degradation demonstrates that protein export and import are mechanistically distinct. J Biol Chem 274:3453-3460

Brown CR, Doxsey SJ, White E, Welch W (1994) Both viral (adenovirus E1B) and cellular (hsp 70, p53) components interact with centrosomes. J Cell Physiol 160:47-60

Cahoreau C, Garnier L, Djiane J, Devauchelle G, Cerutti M (1994) Evidence for N-glycosylation and ubiquitination of the prolactin receptor expressed in a baculovirus-insect cell system. FEBS Lett 350:230-234

Caplan S, Green R, Rocco J, Kurjan J (1991) Glycosylation and structure of the yeast MF α 1 α-factor precursor is important for efficient transport through the secretory pathway. J Bacteriol 173:627-35

Carrano AC, Eytan E, Hershko A, Pagano M (1999) SKP2 is required for ubiquitin-mediated degradation of the CDK inhibitor p27. Nature Cell Biol 1:193-199

Cenciarelli C, Hou D, Hsu KC, Rellahan BL, Wiest DL, Smith HT, Fried VA, Weissman AM (1992) Activation-induced ubiquitination of the T cell antigen receptor. Science 257:795-797

Chang XB, Cui L, Hou YX, Jensen TJ, Aleksandrov AA, Mengos A, Riordan JR (1999) Removal of multiple arginine-framed trafficking signals overcomes misprocessing of delta F508 CFTR present in most patients with cystic fibrosis. Mol Cell 4:137-142

Chen P, Johnson P, Sommer T, Jentsch, S, and Hochstrasser, M (1993) Multiple ubiquitin-conjugating enzymes participate in the in vivo degradation of the yeast MATα 2 repressor Cell 74:357-369

Chillaron J, Haas IG (2000) Dissociation from BiP and retrotranslocation of unassembled immunoglobulin light chains are tightly coupled to proteasome activity. Mol Biol Cell 11:217-226

Ciechanover A (1998) The ubiquitin-proteasome pathway: on protein death and cell life. EMBO J 17:7151-7160

Ciechanover A, Finley D, Varshavsky A (1984) Ubiquitin dependence of selective protein degradation demonstrated in the mammalian cell cycle mutant ts85. Cell 37:57-66

Cook JC, Chock PB (1992) Isoforms of mammalian ubiquitin-activating enzyme. Biol Chem 267:24315-24321

Coux O, Tanaka K, Goldberg AL (1996) Structure and functions of the 20S and 26S proteasomes. Annu Rev Biochem 65:801-847

Crepieux P, Kwon H, Leclerc N, Spencer W, Richard S, Lin R, Hiscott J (1997) I κBα physically interacts with a cytoskeleton-associated protein through its signal response domain. Mol Cell Biol 17:7375-7385

Cummings CJ, Reinstein E, Sun Y, Antalffy B, Jiang Y, Ciechanover A, Orr HT, Beaudet AL, Zoghbi HY (1999) Mutation of the E6-AP ubiquitin ligase reduces nuclear inclusion frequency while accelerating polyglutamine-induced pathology in SCA1 mice. Neuron 24:879-892

Dalemans W, Barbry P, Champigny G, Jallat S, Dott K, Dreyer D, Crystal RG, Pavirani A, Lecocq JP, Lazdunski M (1991) Altered chloride ion channel kinetics associated with the delta F508 cystic fibrosis mutation. Nature 354:526-528

Davarinos NA, Pollenz RS (1999) Aryl hydrocarbon receptor imported into the nucleus following ligand binding is rapidly degraded via the cytosplasmic proteasome following nuclear export. J Biol Chem 274:28708-28715

de Virgilio M, Kitzmuller C, Schwaiger E, Klein M, Kreibich G, Ivessa, NE (1999) Degradation of a short-lived glycoprotein from the lumen of the endoplasmic reticulum: The role of N-linked glycans and the unfolded protein response. Mol Biol Cell 10:4059-4073

de Virgilio M, Weninger H, Ivessa, NE (1998) Ubiquitination is required for the retro-translocation of a short-lived luminal endoplasmic reticulum glycoprotein to the cytosol for degradation by the proteasome. J Biol Chem 273:9734-9743

Deshaies R J (1999) SCF and Cullin/Ring H2-based ubiquitin ligases. Annu Rev Cell Dev Biol 15:435-467

Deshaies RJ, Chau V, Kirschner M (1995) Ubiquitination of the G1 cyclin Cln2p by a Cdc34p-dependent pathway. EMBO J 14:303-312

Desterro JM, Rodriguez MS, Hay RT (1998) SUMO-1 modification of Ikappaßα inhibits NF-kappaß activation. Mol Cell 2:233-239

Diehl JA, Cheng M, Roussel MF, Sherr CJ (1998) Glycogen synthase kinase-3β regulates cyclin D1 proteolysis and subcellular localization. Genes Dev 12:3499-3511

Dohmen RJ, Stappen R, McGrath JP, Forrova H, Kolarov J, Goffeau A, Varshavsky A (1995) An essential yeast gene encoding a homolog of ubiquitin-activating enzyme. J Biol Chem 270:18099-18109

Doye V, Hurt E (1997) From nucleoporins to nuclear pore complexes. Curr Opin Cell Biol 9:401-411

Duprez E, Saurin AJ, Desterro JM, Lallemand-Breitenbach V, Howe K, Boddy MN, Solomon E, de The H, Hay RT, Freemont PS (1999) SUMO-1 modification of the acute promyelocytic leukaemia protein PML: implications for nuclear localisation. J Cell Sci 112:381-393

Dürr G, Strayle J, Plemper R, Elbs S, Klee SK, Catty P, Wolf DH, Rudolph HK (1998) The medial-Golgi ion pump Pmr1 supplies the yeast secretory pathway with Ca2+ and Mn2+ required for glycosylation, sorting, and endoplasmic reticulum-associated protein degradation. Mol Biol Cell 9:1149-1162

Egner R, Kuchler K (1996) The yeast multidrug transporter Pdr5 of the plasma membrane is ubiquitinated prior to endocytosis and degradation in the vacuole. FEBS Lett 378:177-181

Ellgaard L, Molinari M, Helenius A (1999) Setting the standards: Quality control in the secretory pathway. Science 286:1882-1888

Enenkel C, Lehmann A, Kloetzel PM (1998) Subcellular distribution of proteasomes implicates a major location of protein degradation in the nuclear envelope-ER network in yeast. EMBO J 17:6144-6154

Enenkel C, Lehmann A, Kloetzel PM (1999) GFP-labelling of 26S proteasomes in living yeast: insight into proteasomal functions at the nuclear envelope/rough ER. Mol Biol Rep 26:131-135

Fabre E, Hurt E (1997) Yeast genetics to dissect the nuclear pore complex and nucleocytoplasmic trafficking. Annu Rev Genet 31:277-313

Fabunmi RP, Wigley WC, Thomas PJ, DeMartino GN (2000) Activity and regulation of the centrosome-associated proteasome. J Biol Chem 275:409-413

Fang H, Mullins C, Green N (1997) In addition to SEC11, a newly identified gene, SPC3, is essential for signal peptidase activity in the yeast endoplasmic reticulum. J Biol Chem 272:13152-13158

Finger A, Knop M, Wolf, DH (1993) Analysis of two mutated vacuolar proteins reveals a degradation pathway in the endoplasmic reticulum or a related compartment of yeast. Eur J Biochem 218:565-74

Finley D, Ciechanover A, Varshavsky, A (1984) Thermolability of ubiquitin-activating enzyme from the mammalian cell cycle mutant ts85. Cell 37:43-55

Floyd JA, Hamilton BA (1999) Intranuclear inclusions and the ubiquitin-proteasome pathway: digestion of a red herring? Neuron 24:765-766

Freedman DA, Levine AJ (1998) Nuclear export is required for degradation of endogenous p53 by MDM2 and human papillomavirus E6. Mol Cell Biol 18:7288-7293

Freedman DA, Wu L, Levine AJ (1999) Functions of the MDM2 oncoprotein. Cell Mol Life Sci 55:96-107

Galan JM, Cantegrit B, Garnier C, Namy O, Haguenauer-Tsapis R (1998) 'ER degradation' of a mutant yeast plasma membrane protein by the ubiquitin-proteasome pathway. FASEB J 12:315-323

Galan JM, Moreau V, André B, Volland C, Hagenauer-Tsapis R (1996) Ubiquitination mediatedby the npi1p/rsp5p ubiquitin-protein ligase is required for endocytosis of the yeast uracil permease. J Biol Chem 271:10946-10952

Galan JM, Haguenauer-Tsapis R (1997) Ubiquitin lys63 is involved in ubiquitination of a yeast plasma membrane protein. EMBO J 16:5847-5854

Galcheva-Gargova Z, Theroux SJ, Davis RJ (1995) The epidermal growth factor receptor is covalently linked to ubiquitin Oncogene 11:2649-2655

Garcia Mata R, Bebok Z, Sorscher EJ, Sztul ES (1999) Characterization and dynamics of aggresome formation by a cytosolic GFP-chimera. J Cell Biol 146:1239-1254

Gillece P, Luz JM, Lennarz WJ, de la Cruz FJ, Römisch, K (1999) Export of a cysteine-free misfolded secretory protein from the endoplasmic reticulum for degradation requires interaction with protein disulfide isomerase. J Cell Biol 147:1443-1456

Goebl MG, Goetsch L, Byers B (1994) The Ubc3 (Cdc34) ubiquitin-conjugating enzyme is ubiquitinated and phosphorylated in vivo. Mol Cell Biol 14:3022-3029

Goebl MG, Yochem J, Jentsch S, McGrath JP, Varshavsky A, Byers B (1988) The yeast cell cycle gene CDC34 encodes a ubiquitin-conjugating enzyme. Science 241:1331-1335

Gong L, Millas S, Maul GG, Yeh ETH (2000) Differential regulation of sentrinized proteins by a novel sentrin-specific protease J Biol Chem 275:3355-3359

Görlich D, Kutay U (1999) Transport between the cell nucleus and the cytoplasm. Annu Rev Cell Dev Biol 15:607-660

Görlich D, Pante N, Kutay U, Aebi U, Bischoff FR (1996) Identification of different roles for RanGDP and RanGTP in nuclear protein import. EMBO J 15:5584-5594

Görner W, Schüller C, Ruis H (1999) Being at the right place at the right time: the role of nuclear transport in dynamic transcriptional regulation in yeast. Biol Chem 380:147-150

Gostissa M, Hengstermann A, Fogal V, Sandy P, Schwarz SE, Scheffner M, Del Sal G (1999) Activation of p53 by conjugation to the ubiquitin-like protein SUMO-1. EMBO J 18:6462-6471

Govers R, ten Broeke T, van Kerkhof P, Schwartz AL, Strous GJ (1999) Identification of a novel ubiquitin conjugation motif, required for ligand-induced internalization of the growth hormone receptor. EMBO J 18:28-36

Grossi de Sa MF, Martins de Sa C, Harper F, Olink-Coux M, Huesca M, Scherrer K (1988) The association of prosomes with some of the intermediate filament networks of the animal cell. J Cell Biol 107:1517-1530

Hagting A, Jackman M, Simpson K, Pines J (1999) Translocation of cyclin B1 to the nucleus at prophase requires a phosphorylation-dependent nuclear import signal. Curr Biol 9:680-689

Hagting A, Karlsson C, Clute P, Jackman M, Pines J (1998) MPF localization is controlled by nuclear export. EMBO J 17:4127-4138

Hamman BD, Chen JC, Johnson EE, Johnson AE (1997) The aqueous pore through the translocon has a diameter of 40-60 A during cotranslational protein translocation at the ER membrane. Cell 89:535-544

Hampton RY, Rine J (1994) Regulated degradation of HMG-CoA reductase, an integral membrane protein of the endoplasmic reticulum, in yeast. J Cell Biol 125:299-312

Hampton RY, Gardner RG, Rine J (1996) Role of 26S proteasome and HRD genes in the degradation of 3-hydroxy-3-methylglutaryl-CoA reductase, an integral endoplasmic reticulum membrane protein. Mol Biol Cell 7:2029-2044

Handley-Gearhart PM, Stephen AG, Trausch Azar JS, Ciechanover A, Schwartz AL (1994) Human ubiquitin-activating enzyme, E1 Indication of potential nuclear and cytoplasmic subpopulations using epitope-tagged cDNA constructs. J Biol Chem 269:33171-33178

Hazes B, Read RJ (1997) Accumulating evidence suggests that several AB-toxins subvert the endoplasmic reticulum-associated protein degradation pathway to enter target cells. Biochemistry 36:11051-11054

Hein C, Springael JY, Volland C, Haguenauer-Tsapis R, Andre B (1995) NPl1, an essential yeast gene involved in induced degradation of Gap1 and Fur4 permeases, encodes the Rsp5 ubiquitin-protein ligase. Mol Microbiol 18:77-87

Hein WR, Dudler L, Marston WL, Landsverk T, Young AJ, Avila D (1998) Ubiquitination and dimerization of complement receptor type 2 on sheep B cells. J Immunol 161:458-466

Helenius A (1994) How N-linked oligosaccharides affect glycoprotein folding in the endoplasmic reticulum. Mol Biol Cell 5:253-265

Hershko A (1997) Roles of ubiquitin-mediated proteolysis in cell cycle control. Curr Opin Cell Biol 9:788-99

Hershko A, Ciechanover A (1998) The ubiquitin system. Annu Rev Biochem 67:425-479

Hicke L (1999) Gettin' down with ubiquitin: turning off cell-surface receptors, transporters and channels. Trends Cell Biol 9:107-12

Hicke L, Riezman H (1996) Ubiquitination of a yeast plasma membrane receptor signals its ligand-stimulated endocytosis. Cell 84:277-287

Hicke L, Zanolari B, Riezman H (1998) Cytoplasmic tail phosphorylation of the α-factor receptor is required for its ubiquitination and internalization. J Cell Biol 141:349-358

Hill K, Cooper AA (2000) Degradation of unassembled Vph1p reveals novel aspects of the yeast ER quality control system. EMBO J 19:550-561

Hiller MM, Finger A, Schweiger M, Wolf D H (1996) ER degradation of a misfolded luminal protein by the cytosolic ubiquitin-proteasome pathway. Science 273:1725-1728

Hinchcliffe EH, Li C, Thompson EA, Maller JL, Sluder G (1999) Requirement of Cdk2-cyclin E activity for repeated centrosome reproduction in Xenopus egg extracts. Science 283:851-854

Hochstrasser M (1995) Ubiquitin, proteasomes, and the regulation of intracellular protein degradation. Curr Opin Cell Biol 7:215-223

Hochstrasser M (1996) Ubiquitin-dependent protein degradation. Annu Rev Genet 30:405-39

Hochstrasser M (1998) There's the rub: a novel ubiquitin-like modification linked to cell cycle regulation. Genes Dev 12:901-907

Hochstrasser M, Johnson PR, Arendt CS, Amerik AY, Swaminathan S, Swanson R, Li SJ, Laney J, Pals Rylaarsdam R, Nowak J, Connerly PL (1999) The Saccharomyces cerevisiae ubiquitin-proteasome system. Philosophical Transactions of the Royal Society of London Series B Biological Sciences 354:1513-1522

Honda R, Tanaka, H Yasuda H (1997) Oncoprotein MDM2 is a ubiquitin ligase E3 for tumor suppressor p53. FEBS Lett 420:25-27

Hood JK, Silver PA (1999) In or out? Regulating nuclear transport. Curr Opin Cell Biol 11:241-247

Hopper AK (1999) Nucleocytoplasmic transport: Inside out regulation. Curr Biol 9:R803-806

Horak J, Wolf DH (1997) Catabolite inactivation of the galactose transporter in the yeast Saccharomyces cerevisiae: Ubiquitination, endocytosis and degradation in the vaccuole. J Bacteriol 179:1541-1549

Huibregtse JM, Scheffner M, Beaudenon S, Howley PM (1995) A family of proteins structurally and functionally related to the E6-AP ubiquitin-protein ligase. Proc Natl Acad Sci USA 92:2563-2567

Huppa JB, Ploegh HL (1997) The α chain of the T cell antigen receptor is degraded in the cytosol. Immunity 7:113-122

Imamura T, Haruta T, Takata Y, Usui I, Iwata M, Ishihara H, Ishiki M, Ishibashi O, Ueno E, Sasaoka T, Kobayashi M (1998) Involvement of heat shock protein 90 in the degradation of mutant insulin receptors by the proteasome. J Biol Chem 273:11183-11188

Izaurralde E, Kutay U, von Kobbe C, Mattaj IW, Görlich D (1997) The asymmetric distribution of the constituents of the Ran system is essential for transport into and out of the nucleus. EMBO J 16:6535-6547

Jakob CA, Burda P, Roth J, Aebi M (1998) Degradation of misfolded endoplasmic reticulum glycoproteins in Saccharomyces cerevisiae is determined by a specific oligosaccharide structure. J Cell Biol 142:1223-1233

Janson LW, Ragsdale K, Luby-Phelps K (1996) Mechanism and size cutoff for steric exclusion from actin-rich cytoplasmic domains. Biophys J 71:1228-1234

Jeffers M, Taylor GA, Weidner KM, Omura S, Vande Woude GF (1997) Degradation of the Met tyrosine kinase receptor by the ubiquitin-proteasome pathway. Mol Cell Biol 17:799-808

Jensen TJ, Loo MA, Pind S, Williams DB, Goldberg AL, Riordan JR (1995) Multiple proteolytic systems, including the proteasome, contribute to CFTR processing. Cell 83:129-135

Jentsch S (1992) The ubiquitin-conjugation system. Annu Rev Genet 26:179-207

Jentsch S, Seufert W, Sommer T, Reins HA (1990) Ubiquitin-conjugating enzymes: novel regulators of eukaryotic cells. Trends Biochem Sci 15:195-198

Johnson AE, van Waes MA (1999) The translocon: a dynamic gateway at the ER membrane. Annu Rev Cell Dev Biol 15:799-842

Johnson ES, Blobel G (1997) Ubc9p is the conjugating enzyme for the ubiquitin-like protein Smt3p. J Biol Chem 272:26799-26802

Johnson ES, Blobel G (1999) Cell cycle-regulated attachment of the ubiquitin-related protein SUMO to the yeast septins. J Cell Biol 147:981-993

Johnson ES, Schwienhorst I, Dohmen RJ, Blobel G (1997) The ubiquitin-like protein Smt3p is activated for conjugation to other proteins by an Aos1p/Uba2p heterodimer. EMBO J 16:5509-5519

Johnston JA, Ward CL, Kopito RR (1998) Aggresomes: a cellular response to misfolded proteins. J Cell Biol 143:1883-1398

Kaffman A, O'Shea E K (1999) Regulation of nuclear localization: a key to a door. Annu Rev Cell Dev Biol 15:291-339

Kaffman A, Rank NM, O'Neill EM, Huang LS, O'Shea EK (1998) The receptor Msn5 exports the phosphorylated transcription factor Pho4 out of the nucleus. Nature 396:482-486

Kamitani T, Nguyen HP, Kito K, Fukuda-Kamitani, Yeh ET (1998) Covalent modification of the PML by the sentrin family of ubiquitin-like proteins. J Biol Chem 273:3117-3120

Kawahara H, Yokosawa H (1992) Cell cycle-dependent change of proteasome distribution during embryonic development of the ascidian Halocynthia roretzi. Dev Biol 151:27-33

Keller SH, Lindstrom J, Taylor P (1998) Inhibition of glucose trimming with castanospermine reduces calnexin association and promotes proteasome degradation of the α-subunit of the nicotinic acetylcholine receptor. J Biol Chem 273:17064-17072

King RW, Deshaies RJ, Peters JM, Kirschner MW (1996) How proteolysis drives the cell cycle. Science 274:1652-9

Klausner RD, Sitia R (1990) Protein degradation in the endoplasmic reticulum. Cell 62:611-614

Knittler MR, Dirks S, Haas IG (1995) Molecular chaperones involved in protein degradation in the endoplasmic reticulum: quantitative interaction of the heat shock cognate protein BiP with partially folded immunoglobulin light chains that are degraded in the endoplasmic reticulum. Proc Natl Acad Sci USA 92:1764-1768

Knop M, Hauser N, Wolf DH (1996a) N-Glycosylation affects endoplasmic reticulum degradation of a mutated derivative of carboxypeptidase yscY in yeast. Yeast 12:1229-1238

Knop M, Finger A, Braun T, Hellmuth K, Wolf DH (1996b) Der1, a novel protein specifically required for endoplasmic reticulum degradation in yeast. EMBO J 15:753-763

Knuehl C, Seelig A, Brecht B, Henklein P, Kloetzel PM (1996) Functional analysis of eukaryotic 20S proteasome nuclear localization signal. Exp Cell Res 225:67-74

Koegl M, Hoppe T, Schlenker S, Ulrich HD, Mayer TU, Jentsch S (1999) A novel ubiquitination factor, E4, is involved in multiubiquitin chain assembly. Cell 96:635-44

Koepp DM, Harper JW, Elledge SJ (1999) How the cyclin became a cyclin: regulated proteolysis in the cell cycle. Cell 97:431-434

Kolling R, Hollenberg CP (1994) The ABC-transporter Ste6 accumulates in the plasma membrane in a ubiquitinated form in endocytosis mutants. EMBO J 13:3261-3271

Kopito RR (1997) ER Quality Control: The Cytoplasmic Connection. Cell 88:427-430

Kopito RR (1999) Biosynthesis and degradation of CFTR. Physiol Rev 79:167-173

Kornitzer D, Ciechanover A (2000) Modes of regulation of ubiquitin-mediated protein degradation. J Cell Physiol 182:1-11

Kornitzer D, Raboy B, Kulka RG, Fink GR (1994) Regulated degradation of the transcription factor Gcn4. EMBO J 13:6021-6030

Lacey KR, Jackson PK, Stearns T (1999) Cyclin-dependent kinase control of centrosome duplication. Proc Natl Acad Sci USA 96:2817-2822

Lain S, Midgley C, Sparks A, Lane EB, Lane DP (1999) An inhibitor of nuclear export activates the p53 response and induces the localization of HDM2 and p53 to U1A-positive nuclear bodies associated with the PODs. Exp Cell Res 248:457-472

Lain S, Xirodimas D, Lane DP (1999) Accumulating active p53 in the nucleus by inhibition of nuclear export: a novel strategy to promote the p53 tumor suppressor function. Exp Cell Res 253:315-324

Laney JD, Hochstrasser M (1999) Substrate targeting in the ubiquitin system. Cell 97:427-430

Lee DH, Goldberg AL (1996) Selective inhibitors of the proteasome-dependent and vacuolar pathways of protein degradation in Saccharomyces cerevisiae. J Biol Chem 271:27280-27284

Lee GW, Melchior F, Matunis MJ, Mahajan R, Tian Q, Anderson P (1998) Modification of Ran GTPase-activating protein by the small ubiquitin-related modifier SUMO-1 requires Ubc9, an E2-type ubiquitin-conjugating enzyme homologue. J Biol Chem 273:6503-6507

Lehner PJ, Cresswell P (1996) Processing and delivery of peptides presented by MHC class I molecules. Curr Opin Immunol 8:59-67

Lencer WI, Hirst TR, Holmes RK (1999) Membrane traffic and the cellular uptake of cholera toxin. Biochim Biophys Acta 1450:177-190

Leung DW, Spencer SA, Cachianes G, Hammonds RG, Collins C, Henzel WJ, Barnard R, Waters MJ, Wood WI (1987) Growth hormone receptor and serum binding protein: purification, cloning and expression. Nature 330:537-543

Levkowitz G, Waterman H, Ettenberg SA, Katz M, Tsygankov AY, Alroy I, Lavi S, Iwai K, Reiss Y, Ciechanover A, Lipkowitz S, Yarden Y (1999) Ubiquitin ligase activity and tyrosine phosphorylation underlie suppression of growth factor signaling by c-Cbl/Sli-1. Mol Cell 4:1029-1040

Li FN, Johnston M (1997) Grr1 of Saccharomyces cerevisiae is connected to the ubiquitin proteolysis machinery through Skp1: coupling glucose sensing to gene expression and the cell cycle. EMBO J 16:5629-5638

Li SJ, Hochstrasser M (1999) A new protease required for cell-cycle progression in yeast. Nature 398:246-251

Li SJ, Hochstrasser M (2000) The yeast ULP2 (SMT4) gene encodes a novel protease specific for the ubiquitin-like Smt3 protein. Mol Cell Biol 20:2367-2377

Liakopoulos D, Doenges G, Matuschewski K, Jentsch S (1998) A novel protein modification pathway related to the ubiquitin system. EMBO J 17:2208-14

Lisztwan J, Marti A, Sutterluty H, Gstaiger M, Wirbelauer C, Krek W (1998) Association of human CUL-1 and ubiquitin-conjugating enzyme CDC34 with the F-box protein p45(SKP2): evidence for evolutionary conservation in the subunit composition of the CDC34-SCF pathway. EMBO J 17:368-383

Loayza D, Michaelis S (1998) Role for the ubiquitin-proteasome system in the vacuolar degradation of Ste6p, the a-factor transporter in Saccharomyces cerevisiae. Mol Cell Biol 18:779-89

Loeb JD, Schlenstedt G, Pellman D, Kornitzer D, Silver PA, Fink GR (1995) The yeast nuclear import receptor is required for mitosis. Proc Natl Acad Sci USA 92:7647-7651

Loeb KR, Haas AL (1994) Conjugates of ubiquitin cross-reactive protein distribute in a cytoskeletal pattern. Mol Cell Biol 14:8408-8419

Loo MA, Jensen TJ, Cui L, Hou Y, Chang XB, Riordan JR (1998) Perturbation of Hsp90 interaction with nascent CFTR prevents its maturation and accelerates its degradation by the proteasome. EMBO J 17:6879-6887

Lopez-Girona A, Furnari B, Mondesert O, Russell P (1999) Nuclear localization of Cdc25 is regulated by DNA damage and a 14-3-3 protein. Nature 397:172-175

Lucero P, Penalver E, Vela L, Lagunas R (2000) Monoubiquitination is sufficient to signal internalization of the maltose transporter in Saccharomyces cerevisiae. J Bacteriol 182:241-243

Lukacs GL, Mohamed A, Kartner N, Chang XB, Riordan JR, Grinstein S (1994) Conformational maturation of CFTR but not its mutant counterpart (delta F508) occurs in the endoplasmic reticulum and requires ATP. EMBO J 13:6076-6086

Mahajan R, Delphin C, Guan T, Gerace L, Melchior F (1997) A small ubiquitin-related polypeptide involved in targeting RanGAP1 to nuclear pore complex protein RanBP2. Cell 88:97-107

Mahajan R, Gerace L, Melchior F (1998) Molecular characterization of the SUMO-1 modification of RanGAP1 and its role in nuclear envelope association. J Cell Biol 140:259-270

Marchal C, Haguenauer-Tsapis R, Urban-Grimal D (1998) A PEST-like sequence mediates phosphorylation and efficient ubiquitination of yeast uracil permease. Mol Cell Biol 18:314-321

Margottin F, Bour SP, Durand H, Selig L, Benichou S, Richard V, Thomas D, Strebel K, Benarous R (1998) A novel human WD protein, h-β TrCp, that interacts with HIV-1 Vpu connects CD4 to the ER degradation pathway through an F-box motif. Mol Cell 1:565-574

Matunis MJ, Coutavas E, Blobel G (1996) A novel ubiquitin-like modification modulates the partitioning of the Ran-GTPase-activating protein RanGAP1 between the cytosol and the nuclear pore complex. J Cell Biol 135:1457-1470

Matunis MJ, Wu J, Blobel G (1998) SUMO-1 modification and its role in targeting the
    Ran GTPase-activating protein, RanGAP1, to the nuclear pore complex. J Cell
    Biol 140:499-509
Mayer TU, Braun T, Jentsch, S (1998) Role of the proteasome in membrane extrac-
    tion of a short-lived ER-transmembrane protein. EMBO J 17:3251-3257
McCracken AA, Brodsky JL (1996) Assembly of ER-associated protein degradation
    in vitro: dependence on cytosol, calnexin, and ATP. J Cell Biol 132:291-298
McDonald HB, Byers B (1997) A proteasome cap subunit required for spindle pole
    body duplication in yeast. J Cell Biol 137:539-553
McGee TP, Cheng HH, Kumagai H, Omura S, Simoni RD (1996) Degradation of 3-
    hydroxy-3-methylglutaryl-CoA reductase in endoplasmic reticulum membranes
    is accelerated as a result of increased susceptibility to proteolysis. J Biol Chem
    271:25630-25638
Mimnaugh EG, Chavany C, Neckers L (1996) Polyubiquitination and proteasomal
    degradation of the p185c-erbB-2 receptor protein-tyrosine kinase induced by
    geldanamycin. J Biol Chem 271:22796-22801
Miyazawa K, Toyama K, Gotoh A, Hendrie PC, Mantel C, Broxmeyer HE (1994)
    Ligand-dependent polyubiquitination of c-kit gene product: a possible mecha-
    nism of receptor down modulation in M07e cells. Blood 83:137-45
Mori S, Heldin CH, Claesson-Welsh L (1993) Ligand-induced ubiquitination of the
    platelet-derived growth factor β-receptor plays a negative regulatory role in its
    mitogenic signaling. J Biol Chem 268:577-583
Mori S, Claesson-Welsh L, Okuyama Y, Saito Y (1995) Ligand-induced polyubiquiti-
    nation of receptor tyrosine kinases. Biochem Biophys Res Commun 213:32-39
Muller S, Matunis MJ, Dejean A (1998) Conjugation with the ubiquitin-related
    modifier SUMO-1 regulates the partitioning of PML within the nucleus. EMBO J
    17:61-70
Mullins C, Lu Y, Campbell A, Fang H, Green N (1995) A mutation affecting signal
    peptidase inhibits degradation of an abnormal membrane protein in Saccharo-
    myces cerevisiae. J Biol Chem 270:17139-17147
Nakielny S, Dreyfuss G (1999) Transport of proteins and RNAs in and out of the
    nucleus. Cell 99:677-690
Nandi D, Jiang H, Monaco JJ (1996) Identification of MECL-1 (LMP-10) as the third
    IFN-gamma-inducible proteasome subunit. J Immunol 156:2361-2364
Obin MS, Jahngen-Hodge J, Nowell T, Taylor A (1996) Ubiquitinylation and ubiq-
    uitin-dependent proteolysis in vertebrate photoreceptors (rod outer segments)
    Evidence for ubiquitinylation of Gt and rhodopsin. J Biol Chem 271:14473-14484
Otsu M, Urade R, Kito M, Omura F, Kikuchi M (1995) A possible role of ER-60 pro-
    tease in the degradation of misfolded proteins in the endoplasmic reticulum. J
    Biol Chem 270:14958-14961
Pagano M, Tam SW, Theodoras AM, Beer Romero P, Del Sal G, Chau V, Yew PR,
    Draetta GF, Rolfe M (1995) Role of the ubiquitin-proteasome pathway in regulat-
    ing abundance of the cyclin-dependent kinase inhibitor p27. Science 269:682-685
Pal JK, Gounon P, Grossi de Sa MF, Scherrer K (1988) Presence and distribution of
    specific prosome antigens change as a function of embryonic development and
    tissue-type differentiation in Pleurodeles waltl. J Cell Sci 90:555-567
Palmer A, Rivett AJ, Thomson S, Hendil KB, Butcher GW, Fuertes G, Knecht E (1996)
    Subpopulations of proteasomes in rat liver nuclei, microsomes and cytosol. Bio-
    chem J 316:401-407

Pante N, Aebi U (1996) Molecular dissection of the nuclear pore complex. Crit Rev Biochem Mol Biol 31:153-199

Paolini R, Kinet J P (1993) Cell surface control of the multiubiquitination and deubiquitination of high-affinity immunoglobulin E receptors. EMBO J 12:779-786

Papa FR, Amerik AY, Hochstrasser M (1999) Interaction of the Doa4 deubiquitinating enzyme with the yeast 26S proteasome. Mol Biol Cell 10:741-756

Papa FR, Hochstrasser M (1993) The yeast DOA4 gene encodes a deubiquitinating enzyme related to a product of the human tre-2 oncogene. Nature 366:313-319

Peters JM, Franke WW, Kleinschmidt JA (1994) Distinct 19S and 20S subcomplexes of the 26S proteasome and their distribution in the nucleus and the cytoplasm. J Biol Chem 269:7709-7718

Petersen BO, Lukas J, Sorensen CS, Bartek J, Helin K (1999) Phosphorylation of mammalian CDC6 by cyclin A/CDK2 regulates its subcellular localization. EMBO J 18:396-410

Pickart CM (1997) Targeting of substrates to the 26S proteasome. FASEB J 11:1055-1066

Pilon M, Schekman R, Römisch K (1997) Sec61p mediates export of a misfolded secretory protein from the endoplasmic reticulum to the cytosol for degradation. EMBO J 16:4540-4548

Pines J (1999) Cell cycle Checkpoint on the nuclear frontier. Nature 397:104-105

Plemper RK, Böhmler S, Bordallo J, Sommer T, Wolf DH (1997) Mutant analysis links the translocon and BiP to retrograde protein transport for ER degradation. Nature 388:891-895

Plemper RK, Egner R, Kuchler K, Wolf, DH (1998) Endoplasmic reticulum degradation of a mutated ATP-binding cassette transporter Pdr5 proceeds in a concerted action of Sec61 and the proteasome. J Biol Chem 273:32848-32856

Plemper RK, Deak PM, Otto RT, Wolf DH (1999a) Re-entering the translocon from the lumenal side of the endoplasmic reticulum Studies on mutated carboxypeptidase yscY species. FEBS Lett 443:241-245

Plemper RK, Bordallo J, Deak PM, Taxis C, Hitt R, Wolf DH (1999b) Genetic interactions of Hrd3p and Der3p/Hrd1p with Sec61p suggest a retro-translocation complex mediating protein transport for ER degradation. J Cell Sci 112:4123-4134

Plemper RK, Wolf, DH (1999) Retrograde protein translocation: ERADication of secretory proteins in health and disease. Trends Biochem Sci 24:266-270

Plon SE, Leppig KA, Do HN, Groudine M (1993) Cloning of the human homolog of the CDC34 cell cycle gene by complementation in yeast. Proc Natl Acad Sci USA 90:10484-10488

Qu D, Teckman JH, Omura S, Perlmutter DH (1996) Degradation of a mutant secretory protein, α1-antitrypsin Z, in the endoplasmic reticulum requires proteasome activity. J Biol Chem 271:22791-22795

Rapoport TA, Jungnickel B, Kutay U (1996) Protein transport across the eukaryotic endoplasmic reticulum and bacterial inner membranes. Annu Rev Biochem 65:271-303

Reits EAJ, Vos JC, Gromme M, Neefjes J (2000) The major substrates for TAP in vivo are derived from newly synthesized proteins. Nature 13:774-778

Reits EAJ, Benham AM, Plougastel B, Neefjes J, Trowsdale J (1997) Dynamics of proteasome distribution in living cells. EMBO J 16:6087-6094

Rivett AJ (1998) Intracellular distribution of proteasomes. Curr Opin Immunol 10:110-114

Rivett AJ, Mason GG Murray, RZ Reidlinger J (1997) Regulation of proteasome structure and function. Mol Biol Rep 24:99-102

Rivett AJ, Palmer A, Knecht E (1992) Electron microscopic localization of the multi-catalytic proteinase complex in rat liver and in cultured cells. J Histochem Cytochem 40:1165-1172

Roberts BJ, Whitelaw ML (1999) Degradation of the basic helix-loop-helix/Per-ARNT-Sim homology domain dioxin receptor via the ubiquitin/proteasome pathway. J Biol Chem 274:36351-36356

Roth AF, Davis NG (1996) Ubiquitination of the yeast a-factor receptor. J Cell Biol 134:661-674

Roth AF, Sullivan DM, Davis NG (1998) A large PEST-like sequence directs the ubiquitination, endocytosis and vacuolar degradation of the yeast a-factor receptor. J Cell Biol 142:946-961

Roth J, Dobbelstein M, Freedman DA, Shenk T, Levine AJ (1998) Nucleo-cytoplasmic shuttling of the hdm2 oncoprotein regulates the levels of the p53 protein via a pathway used by the human immunodeficiency virus rev protein. EMBO J 17:554-564

Rouillon A, Barbey R, Patton EE, Tyers M, Thomas D (2000) Feedback-regulated degradation of the transcriptional activator Met4 is triggered by the SCF (Met30) complex. EMBO J 19:282-294

Russell SJ, Steger KA, Johnston SA (1999) Subcellular localization, stoichiometry, and protein levels of 26S proteasome subunits in yeast. J Biol Chem 274:21943-21952

Saitoh H, Pu R, Cavenagh M, Dasso M (1997) RanBP2 associates with Ubc9p and a modified form of RanGAP1. Proc Natl Acad Sci USA 94:3736-3741

Sakata N, Stoops JD, Dixon JL (1999) Cytosolic components are required for proteasomal degradation of newly synthesized apolipoprotein B in permeabilized HepG2 cells. J Biol Chem 274:17068-17074

Sato S, Ward CL, Kopito RR (1998) Cotranslational ubiquitination of cystic fibrosis transmembrane conductance regulator in vitro. J Biol Chem 273:7189-7192

Schatz G, Dobberstein B (1996) Common principles of protein translocation across membranes. Science 271:1519-1526

Scheffner M (1999) Moving protein heads for breakdown. Nature 398:103-104

Scheffner M, Huibregtse JM, Vierstra RD, Howley PM (1993) The HPV-16 E6 and E6-AP complex functions as a ubiquitin-protein ligase in the ubiquitination of p53. Cell 75:495-505

Scheffner M, Nuber U, Huibregtse JM (1995) Protein ubiquitination involving an E1-E2-E3 enzyme ubiquitin thioester cascade. Nature 373:81- 83

Schild H, Rammensee, HG (2000) Perfect use of imperfection. Nature 404:709-710

Schmitz A, Herrgen H, Winkler A, Herzog V (2000) Cholera toxin is exported from microsomes by the Sec61 complex. J Cell Biol 148:1203-1212

Schubert U, Anton LC, Gibbs J, Norbury CC, Yewdell JW, Bennink JR (2000) Rapid degradation of a large fraction of newly synthesized proteins by proteasomes. Nature 404:770-774

Schubert U, Anton LC, Bacik I, Cox JH, Bour S, Bennink JR, Orlowski M, Strebel K, Yewdell JW (1998) CD4 glycoprotein degradation induced by human immunodeficiency virus type 1 Vpu protein requires the function of proteasomes and the ubiquitin-conjugating pathway. J Virol 72:2280-2288

Schüle T, Rose M, Entian KD, Thumm M, Wolf DH (2000) Ubc8p functions in catabolite degradation of fructose-1,6-bisphosphatase in yeast. EMBO J 19:2161-2167

Schwartz AL, Ciechanover A (1999) The ubiquitin-proteasome pathway and pathogenesis of human diseases. Annu Rev Med 50:57-74

Schwarz SE, Matuschewski K, Liakopoulos D, Scheffner M, Jentsch S (1998) The ubiquitin-like proteins SMT3 and SUMO-1 are conjugated by the UBC9 E2 enzyme. Proc Natl Acad Sci USA 95:560-564

Scidmore MA, Okamura HH, Rose MD (1993) Genetic interactions between KAR2 and SEC63, encoding eukaryotic homologues of DnaK and DnaJ in the endoplasmic reticulum. Mol Biol Cell 4:1145-1159

Seufert W, Futcher B, Jentsch S (1995) Role of a ubiquitin-conjugating enzyme in degradation of S- and M-phase cyclins. Nature 373:78-81

Shamu CE, Story CM, Rapoport TA, Ploegh HL (1999) The pathway of US11-dependent degradation of MHC class I heavy chains involves a ubiquitin-conjugated intermediate. J Cell Biol 147:45-57

Shih SC, Sloper-Mould KE, Hicke L (2000) Monoubiquitin carries a novel internalization signal that is appended to activated receptors. EMBO J 19:187-198

Shimkets RA, Lifton RP, Canessa CM (1997) The activity of the epithelial sodium channel is regulated by clathrin-mediated endocytosis. J Biol Chem 272:25537-25541

Silver ET, Gwozd TJ, Ptak C, Goebl M, Ellison MJ (1992) A chimeric ubiquitin conjugating enzyme that combines the cell cycle properties of CDC34 (UBC3) and the DNA repair properties of RAD6 (UBC2): implications for the structure, function and evolution of the E2s. EMBO J 11:3091-3098

Simpson JC, Roberts LM, Romisch K, Davey J, Wolf DH, Lord JM (1999) Ricin A chain utilises the endoplasmic reticulum-associated protein degradation pathway to enter the cytosol of yeast. FEBS Lett 459:80-84

Smart P, Lane EB, Lane DP, Midgley C, Vojtesek B, Lain S (1999) Effects on normal fibroblasts and neuroblastoma cells of the activation of the p53 response by the nuclear export inhibitor leptomycin B. Oncogene 18:7378-7386

Snyder PM, Price MP, McDonald FJ, Adams CM, Volk KA, Zeiher BG, Stokes JB, Welsh MJ (1995) Mechanism by which Liddle's syndrome mutations increase activity of a human epithelial Na+ channel. Cell 83:969-978

Sommer T, Jentsch S (1993) A protein translocation defect linked to ubiquitin conjugation at the endoplasmic reticulum. Nature 365:176-179

Sommer T, Wolf DH (1997) Endoplasmic reticulum degradation: reverse protein flow of no return. FASEB J 11:1227-1233

Song A, Wang Q, Goebl MG, Harrington MA (1998) Phosphorylation of nuclear MyoD is required for its rapid degradation. Mol Cell Biol 18:4994-4999

Springael JY, Andre B (1998) Nitrogen-regulated ubiquitination of the Gap1 permease of Saccharomyces cerevisiae. Mol Biol Cell 9:1253-1263

Staub O, Dho S, Henry P, Correa J, Ishikawa T, McGlade J, Rotin D (1996) WW domains of Nedd4 bind to the proline-rich PY motifs in the epithelial Na+ channel deleted in Liddle's syndrome. EMBO J 15:2371-2380

Staub O, Gautschi I, Ishikawa T, Breitschopf K, Ciechanover A, Schild L, Rotin D (1997) Regulation of stability and function of the epithelial Na+ channel (ENaC) by ubiquitination. EMBO J 16:6325-6336

Stephen AG, Trausch Azar JS, Ciechanover A, Schwartz AL (1996) The ubiquitin-activating enzyme E1 is phosphorylated and localized to the nucleus in a cell cycle-dependent manner. J Biol Chem 271:15608-15614

Stephen AG, Trausch Azar JS, Handley Gearhart PM, Ciechanover A, Schwartz AL (1997) Identification of a region within the ubiquitin-activating enzyme required for nuclear targeting and phosphorylation. J Biol Chem 272:10895-10903

Sternsdorf T, Jensen K, Will H (1997) Evidence for covalent modification of the nuclear dot-associated proteins PML and Sp100 by PIC1/SUMO-1. J Cell Biol 139:1621-1634

Stommel JM, Marchenko ND, Jimenez GS, Moll UM, Hope TJ, Wahl GM (1999) A leucine-rich nuclear export signal in the p53 tetramerization domain: regulation of subcellular localization and p53 activity by NES masking. EMBO J 18:1660-1672

Strous GJ, Govers R (1999) The ubiquitin-proteasome system and endocytosis. J Cell Sci 112:1417-1423

Strous GJ, van Kerkhof P, Govers R, Ciechanover A, Schwartz AL (1996) The ubiquitin conjugation system is required for ligand-induced endocytosis and degradation of the growth hormone receptor. EMBO J 15:3806-3812

Su K, Stoller T, Rocco J, Zemsky J, Green R (1993) Pre-Golgi degradation of yeast prepro-α-factor expressed in a mammalian cell Influence of cell type-specific oligosaccharide processing on intracellular fate. J Biol Chem 268:14301-14309

Sutterlüty H, Chatelain E, Marti A, Wirbelauer C, Senften M, Muller U, Krek W (1999) p45SKP2 promotes p27^{Kip1} degradation and induces S phase in quiescent cells. Nat Cell Biol 1:207-214

Suzuki T, Seko A, Kitajima K, Inoue Y, Inoue S (1994) Purification and enzymatic properties of peptide:N-glycanase from C3H mouse-derived L-929 fibroblast cells Possible widespread occurrence of post-translational remodification of proteins by N-deglycosylation. J Biol Chem 269:17611-17618

Swaminathan S, Amerik AY, Hochstrasser M (1999) The Doa4 deubiquitinating enzyme is required for ubiquitin homeostasis in yeast. Mol Biol Cell 10:2583-2594

Tanaka K, Yoshimura T, Tamura T, Fujiwara T, Kumatori A, Ichihara A (1990) Possible mechanism of nuclear translocation of proteasomes. FEBS Lett 271:41-46

Tao W, Levine AJ (1999) Nucleocytoplasmic shuttling of oncoprotein Hdm2 is required for Hdm2-mediated degradation of p53. Proc Natl Acad Sci USA 96:3077-3080

Terrell J, Shih S, Dunn R, Hicke L (1998) A function for monoubiquitination in the internalization of a G protein-coupled receptor. Mol Cell 1:193-202

Tomoda K, Kubota Y, Kato J (1999) Degradation of the cyclin-dependent-kinase inhibitor p27^{Kip1} is instigated by Jab1. Nature 398:160-165

Toyoshima F, Moriguchi T, Wada A, Fukuda M, Nishida, E (1998) Nuclear export of cyclin B1 and its possible role in the DNA damage-induced G2 checkpoint. EMBO J 17:2728-2735

Tsvetkov LM, Yeh KH, Lee SJ, Sun H, Zhang H (1999) p27(^{Kip1}) ubiquitination and degradation is regulated by the SCF(Skp2) complex through phosphorylated Thr187 in p27. Curr Biol 9:661-664

van Kerkhof P, Govers R, Alves dos Santos CM, Strous GJ (2000) Endocytosis and degradation of the growth hormone receptor are proteasome-dependent. J Biol Chem 275:1575-1580

Vieira AV, Lamaze C, Schmid SL (1996) Control of EGF receptor signaling by clathrin-mediated endocytosis. Science 274:2086-2089

Vijay-Kumar S, Bugg CE, Wilkinson KD, Cook WJ (1985) Three-dimensional structure of ubiquitin at 28 A resolution. Proc Natl Acad Sci USA 82:3582-3585

Wang HR, Kania M, Baumeister W, Nederlof PM (1997) Import of human and Thermoplasma 20S proteasomes into nuclei of HeLa cells requires functional NLS sequences. Eur J Cell Biol 73:105-113

Ward CL, Kopito RR (1994) Intracellular turnover of cystic fibrosis transmembrane conductance regulator Inefficient processing and rapid degradation of wild-type and mutant proteins. J Biol Chem 269:25710-25718

Ward CL, Omura S, Kopito RR (1995) Degradation of CFTR by the ubiquitin-proteasome pathway. Cell 83:121-127

Werner ED, Brodsky JL, McCracken AA (1996) Proteasome-dependent endoplasmic reticulum-associated protein degradation: an unconventional route to a familiar fate. Proc Natl Acad Sci USA 93:13797-13801

Wesche J, Rapak A, Olsnes S (1999) Dependence of ricin toxicity on translocation of the toxin A-chain from the endoplasmic reticulum to the cytosol. J Biol Chem 274:34443-34449

Wiertz EJ, Jones TR, Sun L, Bogyo M, Geuze HJ, Ploegh HL (1996a) The human cytomegalovirus US11 gene product dislocates MHC class I heavy chains from the endoplasmic reticulum to the cytosol. Cell 84:769-779

Wiertz EJ, Tortorella D, Bogyo M, Yu J, Mothes W, Jones TR, Rapoport TA, Ploegh HL (1996b) Sec61-mediated transfer of a membrane protein from the endoplasmic reticulum to the proteasome for destruction. Nature 384:432-438

Wigley WC, Fabunmi RP, Lee MG, Marino CR, Muallem S, DeMartino GN, Thomas PJ (1999) Dynamic association of proteasomal machinery with the centrosome. J Cell Biol 145:481-490

Wilkinson BM, Tyson JR, Reid PJ, Stirling CJ (2000) Distinct domains within yeast Sec61p involved in post-translational translocation and protein dislocation. J Biol Chem 275:521-529

Wilkinson CR, Wallace M, Morphew M, Perry P, Allshire R, Javerzat JP, McIntosh JR, Gordon C (1998) Localization of the 26S proteasome during mitosis and meiosis in fission yeast. EMBO J 17:6465-6476

Wilkinson KD (1997) Regulation of ubiquitin-dependent processes by deubiquitinating enzymes. FASEB J 11:1245-1256

Winston JT, Koepp DM, Zhu C, Elledge SJ, Harper JW (1999) A family of mammalian F-box proteins. Curr Biol 9:1180-1182

Xie Y, Varshavsky A (1999) The E2-E3 interaction in the N-end rule pathway: the RING-H2 finger of E3 is required for the synthesis of multiubiquitin chain. EMBO J 18:6832-6844

Xiong X, Chong E, Skach W R (1999) Evidence that endoplasmic reticulum (ER)-associated degradation of cystic fibrosis transmembrane conductance regulator is linked to retrograde translocation from the ER membrane. J Biol Chem 274:2616-2624

Yan C, Lee LH, Davis LI (1998) Crm1p mediates regulated nuclear export of a yeast AP-1-like transcription factor. EMBO J 17:7416-7429

Yang J, Bardes ES, Moore JD, Brennan J, Powers MA, Kornbluth S (1998) Control of cyclin B1 localization through regulated binding of the nuclear export factor CRM1. Genes Dev 12:2131-1243

Yang M, Omura, S Bonifacino JS, Weissman AM (1998) Novel aspects of degradation of T cell receptor subunits from the endoplasmic reticulum (ER) in T cells: importance of oligosaccharide processing, ubiquitination, and proteasome-dependent removal from ER membranes. J Exp Med 187:835-846

Yang Y, Fruh K, Ahn K, Peterson PA (1995) In vivo assembly of the proteasomal complexes, implications for antigen processing. J Biol Chem 270:27687-27694

Yarden Y, Escobedo JA, Kuang WJ, Yang-Feng TL, Daniel TO, Tremble PM, Chen EY, Ando ME, Harkins RN, Francke U, et al (1986) Structure of the receptor for platelet-derived growth factor helps define a family of closely related growth factor receptors. Nature 323:226-232

Yu H, Kopito RR (1999) The role of multiubiquitination in dislocation and degradation of the α subunit of the T cell antigen receptor. J Biol Chem 274:36852-36858

Yuk MH, Lodish HF (1993) Two pathways for the degradation of the H2 subunit of the asialoglycoprotein receptor in the endoplasmic reticulum. J Cell Biol 123:1735-1749

Zhou MY, Schekman R (1999) The engagement of Sec61p in the ER dislocation process. Mol Cell 4:925-934

Transgenic Models of α_2-Adrenergic Receptor Subtype Function

L. Hein

Institut für Pharmakologie und Toxikologie, Universität Würzburg,
Versbacher Straße 9, 97078 Würzburg, Germany

Contents

Abbreviations

α_{2A}-KO, α_{2A}-adrenergic receptor knockout; α_{2A}-D79N, aspartic acid 79 to asparagine mutation of the α_{2A}-adrenergic receptor; ERK, extracellular signal-regulated kinase; G_i, inhibitory G protein; GRK, G protein-coupled receptor kinase; G_s, stimulatory G protein; MAPK, mitogen-activated protein kinase; N_2O, nitrous oxide; PC12 cells, rat pheochromocytoma cell line; WT, wild-type

1 Introduction

Adrenergic receptors are located throughout the body on neuronal and non-neuronal cells where they mediate a diverse range of responses to the endogenous catecholamines adrenaline and noradrenaline. To date, nine

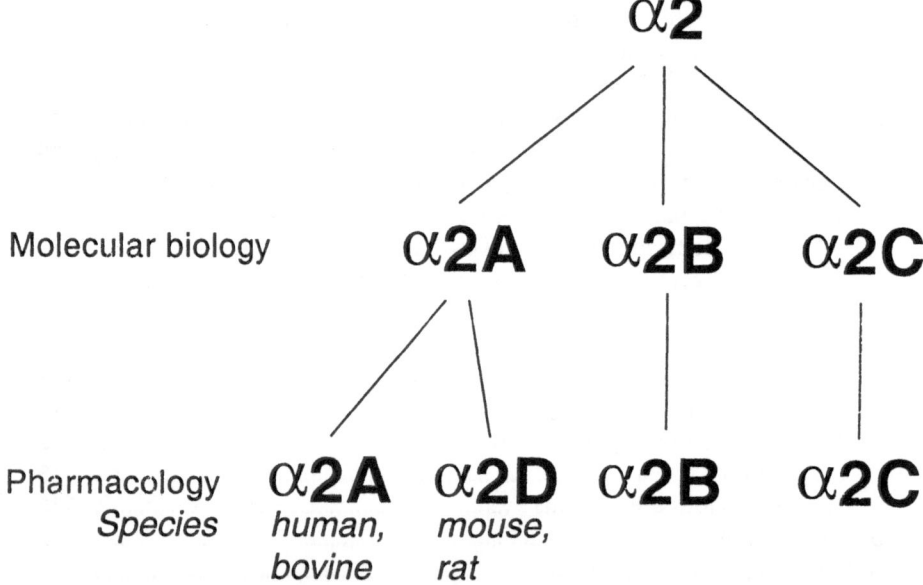

Fig. 1. Overview of α_2-adrenergic receptor subtype nomenclature. In many species, three genes encoding for distinct α_2-adrenergic receptor subtypes have been identified. These receptors are termed α_{2A}, α_{2B}, and α_{2C}. Pharmacological experiments with subtype-preferring antagonists have led to a further subdivision into α_{2A}- and α_{2D}-receptors (Bylund et al. 1994). These two receptors are species variants of the same gene, with the pharmacological α_{2A}-subtype being expressed in human and bovine tissues and the α_{2D}-subtype in mouse and rat. Point mutation of a serine residue in the fifth transmembrane domain of the α_{2A}-receptor to alanin confers decreased affinity to rauwolscine and yohimbine and thus constitutes α_{2D}-pharmacology (Link et al. 1992)

adrenergic receptors have been cloned. Based on pharmacological and molecular biology nomenclature, the adrenergic receptor family can be subdivided into three α_1-receptors (α_{1A}, α_{1B}, α_{1D}), three α_2-receptors (α_{2A}, α_{2B}, α_{2C}) and three β-adrenergic receptors (β_1, β_2, β_3). Physiological functions of α_1- and β- receptor subtypes have been reviewed extensively elsewhere (Docherty 1998; Lowell 1998; Rohrer 1998; Rohrer and Kobilka 1998; Dzimiri 1999; Liggett 1999; Rohrer, 2000). This review focusses on the significance of α_2-receptor subtype diversity.

1.1 α_2-Receptor Subtype Nomenclature

Three genes encoding for α_2-adrenergic receptor subtypes have been identified from several species, termed α_{2A}, α_{2B}, and α_{2C}, respectively (Fig. 1, 2).

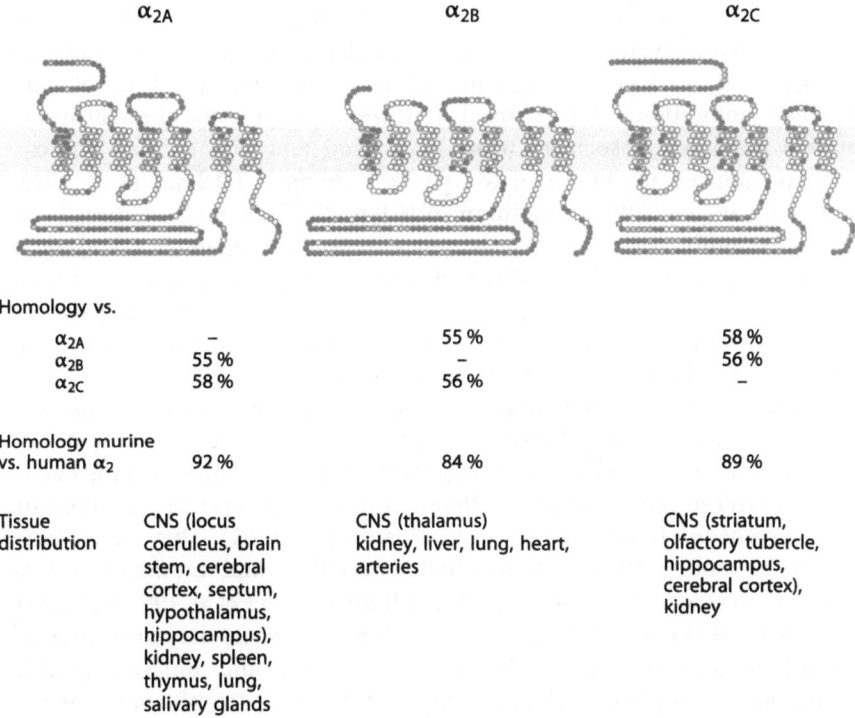

	α_{2A}		α_{2B}	α_{2C}
Homology vs.				
α_{2A}	–		55 %	58 %
α_{2B}	55 %		–	56 %
α_{2C}	58 %		56 %	–
Homology murine vs. human α_2	92 %		84 %	89 %
Tissue distribution	CNS (locus coeruleus, brain stem, cerebral cortex, septum, hypothalamus, hippocampus), kidney, spleen, thymus, lung, salivary glands		CNS (thalamus) kidney, liver, lung, heart, arteries	CNS (striatum, olfactory tubercle, hippocampus, cerebral cortex), kidney

Fig. 2. α_2-Adrenergic receptor subtypes. The putative membrane topology of the three mouse α_2-receptor subtypes is schematically depicted. Open circles represent amino acids which are identical between all subtypes, gray circles represent amino acids which are conserved among two subtypes, dark circles are non-identical amino acids. Tissue distribution of α_2-receptor subtypes is based on mRNA and protein expression; references are given in the text

The pharmacological profile of the α_{2A}-subtype differs significantly between species, thus giving rise to the pharmacological subtypes α_{2A} in humans, rabbits and pigs and α_{2D} in rats, mice and guinea pigs (Bylund et al. 1994). Part of the pharmacological difference between α_{2A}- and α_{2D}-receptors can be explained by a Ser-Ala mutation in the fifth transmembrane helix of the α_{2A}-receptor rendering this receptor less sensitive to rauwolscine and yohimbine binding (Link et al. 1992). For the purpose of this review, the genetic nomenclature of α_{2A}, α_{2B}, and α_{2C} receptor subtypes will be used.

1.2 Intracellular Signal Transduction

α_2-Adrenergic receptors can regulate a wide range of signalling pathways via interaction with multiple heterotrimeric G_i proteins ($G\alpha_{i1}$, $G\alpha_{i2}$, $G\alpha_{i3}$) including inhibition of adenylyl cyclase (Cotecchia et al. 1990; Wise et al. 1997), stimulation of phospholipase D (MacNulty et al. 1992), stimulation of ERK/mitogen-activated protein kinases (Alblas et al. 1993; Anderson and Milligan 1994; Koch et al. 1994), stimulation of K^+ currents (Surprenant et al. 1992) and inhibition of Ca^{2+} currents (Surprenant et al. 1992). In neurons, G protein $\beta\gamma$-dimers associated with $G\alpha_{OA}$, $G\alpha_{OB}$ and $G\alpha_{i2}$ mediate the α_2-receptor-induced inhibition of N-type Ca^{2+} channels (Delmas et al. 1999; Jeong and Ikeda 2000). In addition to N-type Ca^{2+} channels, neuronal α_2-receptors inhibit voltage-gated P- and Q-type Ca^{2+} channels (Waterman 1997). Furthermore, at least with high receptor expression levels and high receptor occupancy, effector regulation via activation of other G proteins can be uncovered (Eason et al. 1992; Pepperl and Regan 1993; Eason et al. 1994; Eason and Liggett 1995; Pierce et al. 2000).

In the α_{2A}-receptor, mutation of an aspartic acid residue (Asp^{79}) which is highly conserved among G protein-coupled receptors to asparagine (D79N) led to selective uncoupling of the α_{2A}-D79N receptor from K^+ channel activation in AtT20 cells without an alteration in adenylyl cyclase inhibition or Ca^{2+} current inhibition (Fig. 3) (Surprenant et al. 1992). This aspartate is conserved among all G protein-coupled receptors and mutation of this aspartate to asparagine reduces agonist affinity (Strader et al. 1987; Chung et al. 1988; Fraser et al. 1989; Wang et al. 1991) and prevents the modulation of agonist binding by cations (Guyer et al. 1990) and nonhydrolyzable guanosine triphosphate analogs (Chung et al. 1988). Recombinant fusions of the α_{2A}-D79N receptor with pertussis toxin-resistant forms of the G proteins $G\alpha_{i1}$, $G\alpha_{i2}$, or $G\alpha_{i3}$ have demonstrated that point mutation of the aspartic acid residue non-selectively disrupted the capacity of the α_{2A}-receptor to activate these closely related G proteins (Ward and Milligan 1999). Recent studies have also demonstrated that the α_{2A}-D79N mutation is conforma-

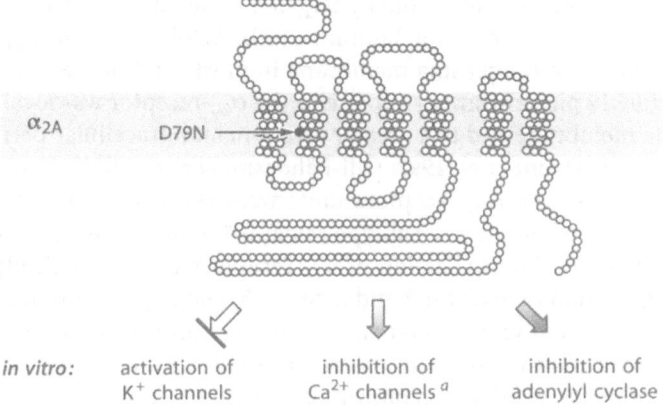

in vitro: activation of   inhibition of   inhibition of
                  K^+ channels Ca^{2+} channels [a]   adenylyl cyclase

in vivo: decreased α_{2A}-D79N receptor expression (20 % of WT)

Fig. 3. Signal transduction of the α_{2A}-D79N adrenergic receptor. Mutation of the aspartic acid residue 79 in the second transmembrane domain of the mouse α_{2A}-adrenergic receptor to asparagine (D79N) selectively uncoupled this receptor mutant from activation of K^+ currents in vitro without interfering with Ca^{2+} channel or adenylyl cyclase inhibition (Surprenant et al. 1992). In vivo, the α_{2A}-D79N receptor was expressed at 20% of the level of the wild-type α_{2A}-receptor (MacMillan et al. 1996), and Ca^{2+} channel inhibition was also blunted (*a*) (Lakhlani et al. 1997)

tionally unstable and readily turns over on the cell surface (Wilson and Limbird, 2000). The observed phenotype of the α_{2A}-D79N mutation may thus be due to differential levels of signal amplification required to produce K^+ channel activation, Ca^{2+} channel and adenylyl cyclase inhibition in different cell types or due to differences in G protein expression levels or localization and/or to intrinsic differences in receptor stability and turnover.

Upon stimulation with agonist, α_{2A}-adrenergic receptors undergo subtype-specific, short-term desensitization (Eason and Liggett 1992; Kurose and Lefkowitz 1994). Desensitization of α_{2A}- and α_{2B}-receptors is mediated by phosphorylation of intracellular receptor domains by specific G protein-coupled receptor kinases type 2 and 3 (GRK2 and GRK3; Jewell-Motz and Liggett 1996; Wu et al. 1997). No phosphorylation of the α_{2C}-subtype by GRKs could be detected (Eason and Liggett 1992; Jewell-Motz and Liggett 1996). In neurons, inhibition of voltage-dependent Ca^{2+} channels mediated by α_2-adrenergic receptors desensitizes slowly with prolonged exposure of transmitter and desensitization is mediated by GRK3 kinase (Diverse-Pierluissi et al. 1996).

α_2-Adrenergic receptors are differentially targeted to membrane domains in cells (for review see Saunders and Limbird 1999). While α_{2A}- and α_{2B}-receptors were targeted to the plasma membrane in most cell lines, including differentiated PC12 pheochromocytoma cells, the α_{2C}-receptor was localized at the plasma membrane and to a greater extent in an intracellular perinuclear compartment (Daunt et al. 1997; Olli-Lähdesmäki et al. 1999). Upon agonist stimulation, α_{2B}- and α_{2C}-receptors underwent reversible internalization into endosomes, while the α_{2A}-subtype remained in the plasma membrane (Daunt et al. 1997). Internalization of α_{2B}-receptors was dramatically enhanced by coexpression of arrestin-2 and arrestin-3, and redistribution of receptors to clathrin-coated vesicles and endosomes was observed (DeGraff et al. 1999). Internalization of α_{2B}- and α_{2C}-receptors was inhibited by coexpression of dominant negative dynamin-K44A (DeGraff et al. 1999); for the α_{2B}-receptor, coexpression of dynamin-K44A gave variable results (DeGraff et al. 1999; Schramm and Limbird 1999). Endocytosis did not appear to be required for α_2-adrenergic receptor-mediated p42/p44 MAP kinase activation (DeGraff et al. 1999; Schramm and Limbird 1999).

1.3 α_2-Receptor Tissue Distribution

The three α_2-receptor subtypes have unique patterns of tissue distribution in the central nervous system and in peripheral tissues (Fig. 2) (Nicholas et al. 1993; Aoki et al. 1994; Scheinin et al. 1994; MacDonald and Scheinin 1995; Nicholas et al. 1996; Rosin et al. 1996; Talley et al. 1996; Uhlen et al. 1997). These studies argue against complete functional redundancy for these receptor subtypes. The α_{2C}-receptor appears to be expressed primarily in the central nervous system (striatum, olfactory tubercle, hippocampus and cerebral cortex), although very low levels of its mRNA are present in the kidney. In contrast, the α_{2B}-receptor shows primarily peripheral expression (kidney, liver, lung, heart) and only low level expression in thalamic nuclei. The α_{2A}-receptor is expressed more widely throughout the central nervous system, including the locus coeruleus, brain stem nuclei, cerebral cortex, septum, hypothalamus, hippocampus) and in the periphery (kidney, spleen, thymus, lung, salivary gland).

In mice, expression of α_2-receptors can be detected during early embryonic development, starting at day 9.5 postcoitus for the α_{2A}-receptor, day 11.5 for the α_{2B}-, and 14.5 for the α_{2C}-subtype (Wang and Limbird 1997). α_{2A}-receptor mRNA showed a widespread distribution in the developing embryo, including the developing stomach and cecum, craniofacial regions and central nervous system. Also, α_{2A}-mRNA was expressed in the interdigital mesenchyme in parallel with digital separation (Wang and Limbird

1997). The α_{2B}-receptor could be detected in the developing liver when the liver is the principal site for hematopoiesis. α_{2C}-receptor mRNA was found in the nasal cavity and cerebellar primordium. The functional role of α_2-receptors for embryonic development is unclear at present. However, the α_{2A}-subtype can induce apoptosis in vitro (Wang and Limbird 1997).

2 Transgenic Mouse Models

Using molecular genetics and transgenic techniques, several mouse lines overexpressing or lacking α_2-adrenergic receptor subtypes have been generated (for review, see MacDonald et al. 1997; Kable et al. 2000). The genes encoding for the three murine α_2-receptors were disrupted in embryonic stem cells and knockout mouse lines lacking α_{2A}-, α_{2B}-, and α_{2C}-receptors were established (Link et al. 1995; Link et al. 1996; Altman et al. 1999). Mice with homozygous deletions in single α_2-receptor genes were viable and appeared grossly normal. From heterozygous crosses, homozygous α_{2A}-KOs and α_{2C}-KOs were born at the expected Mendelian ratios, however, α_{2B}-KOs were less abundant in these crosses than expected (Link et al. 1996). Thus, α_2-receptors may play an important role during embryogenesis and development. As the three α_2-receptor genes are localized on different chromosomes, knockout mouse lines could be crossed to generate strains lacking two α_2-receptor subtypes. So far, only mice lacking α_{2A}- and α_{2C}-receptors (α_{2AC}-KO) could be recovered from these crosses (Hein et al. 1999). All other combinations, i.e. α_{2AB}-KO and α_{2BC}-KO, seemed to be lethal during intrauterine development, further emphazising the significance of α_2-adrenergic receptors for embryonic development.

In order to investigate the in vivo significance of individual signalling pathways of the α_{2A}-subtype, Limbird and colleagues used gene-targeting to mutate the α_{2A}-receptor gene to express an Asp79->Asn (D79N) α_{2A}-receptor in mice (Fig. 3) (MacMillan et al. 1996). The D79N mutation substitutes asparagine for an aspartate residue at position 79, which is predicted to lie within the second transmembrane region of the α_{2A}-receptor and is highly conserved among G protein-coupled receptors (Fig. 2). In vitro, the α_{2A}-D79N receptor was found to be selectively uncoupled from activation of K^+ currents, but remained coupled to inhibition of voltage-gated Ca^{2+} channels and of cAMP production characteristic for the wild-type α_{2A}-receptor (Surprenant et al. 1992). Similar to the α_{2A}-KO mice, mice carrying the α_{2A}-D79N mutation were viable and developed normally (MacMillan et al. 1996). Surprisingly, the density of α_{2A}-D79N receptors was greatly reduced by 80% in brain membranes of homozygous transgenic mice, although there was no change in the abundance of mRNA encoding for α_{2A}-D79N compared with the wild-type mRNA (MacMillan et al. 1996). These results indicate that, in vivo, the α_{2A}-D79N receptor may be improperly processed or stabilized in its target cells (MacMillan et al. 1998). In fact, subsequent studies suggest that the diminished steady state density of the α_{2A}-D79N receptor is due to accelerated turnover of this conformationally unstable mutant (Wilson and Limbird 2000). As expected from in vitro experiments, the α_{2A}-D79N receptor was uncoupled from K^+ channel activation in locus coeruleus neurons (Lakhlani et al. 1997). However, the inhibitory effect of α_2-agonists on Ca^{2+} channels in superior cervical ganglion cells was also blunted while the agonist-dependent inhibition of adenylyl cyclase appeared to be normal in these mice (Lakhlani et al. 1997). Thus for most, but not all, physiological function of α_2-receptors tested, the α_{2A}-D79N mutation turned out be behave like a "functional knockout" (MacMillan et al. 1996).

In addition to these gene-targeted mouse models, mouse lines overexpressing the α_{2C}-adrenergic receptor under control of its own promoter were generated by Kobilka and colleagues (Sallinen et al. 1997). The promotor elements present in the α_{2C}-transgene construct were capable of directing the increased receptor expression to those brain regions, and probably also to those cells, which normally express endogenous α_{2C}-receptors. In these transgenic mice, 3-fold overexpression of the α_{2C}-receptor was detected in caudate-putamen and in the stratum radiatum of the CA1 region of the hippocampus (Sallinen et al. 1997).

3 Presynaptic α_2-Adrenergic Receptors

Prejunctional α_2-receptors on adrenergic nerves mediate a negative feedback whereby released noradrenaline inhibits its own further release (for reviews, see Langer 1974; Starke et al. 1975; Westfall 1977; Starke et al. 1989; Langer 1997). Using a series of antagonists, measurements of ^3H-noradrenaline release in isolated tissues had suggested that presynaptic α_2-receptors belong – at least predominantly – to the pharmacological α_{2A}- or α_{2D}-subtypes (for review, see (Trendelenburg et al. 1993). This finding was confirmed in mice lacking α_{2A}-receptors by two experimental approaches: first, inhibition of the electrically evoked twitch response of isolated vasa deferentia by exogenous α_2-agonists was tested (Altman et al. 1999), second, the stimulated release of ^3H-noradrenaline from isolated tissues was measured (Fig. 4) (Hein et al. 1999; Trendelenburg et al. 1999). With both methods, the maximal inhibition of presynaptic transmitter release by a non-subtype selective α_2-agonist was decreased in mice lacking α_{2A}-receptors but not in lines lacking α_{2B}- or α_{2C}-receptors. These data indicate,

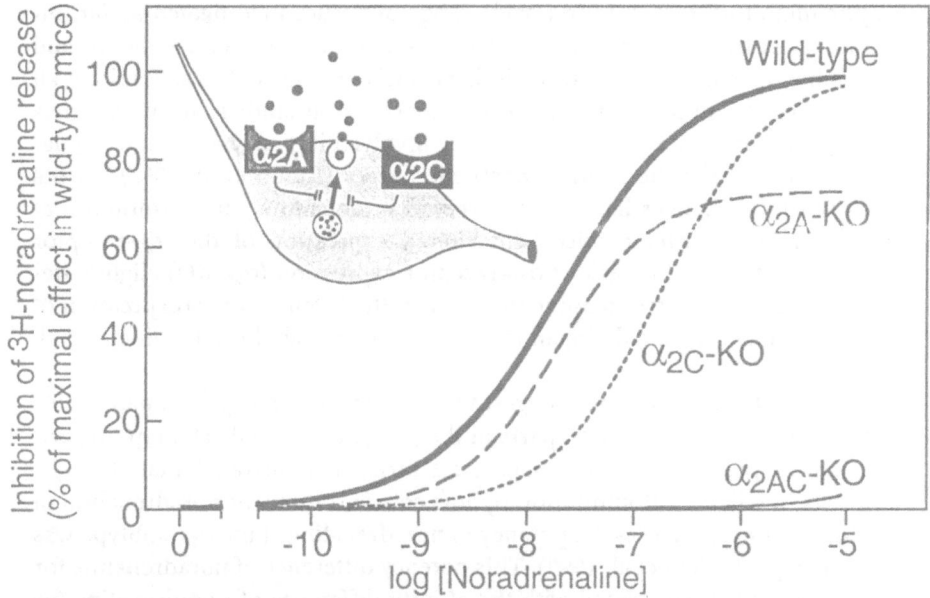

Fig. 4. Inhibition of noradrenaline release by α_{2A}- and α_{2C}-receptors. In atria from wild-type mice, exogenous noradrenaline completely inhibited release of ^{3}H-noradrenaline from sympathetic nerves. After deletion of the α_{2A}-receptor gene, the maximal inhibitory effect of noradrenaline was decreased, and knockout of the α_{2C}-receptor gene led to a rightward shift of the noradrenaline concentration response curve. Only in atria from mice lacking α_{2A}- and α_{2C}-receptors (α_{2AC}-KO) was the inhibitory effect of noradrenaline completely abolished, indicating that α_{2A}- and α_{2C}-adrenergic receptors are both required for presynaptic feedback inhibition of transmitter release (adapted from Hein et al. 1999)

that indeed the α_{2A}-subtype is the main inhibitory presynaptic autoreceptor on adrenergic neurons. However, in all tissues investigated from α_{2A}-KO mice α_2-agonists could still inhibit the release of noradrenaline (Altman et al. 1999; Trendelenburg et al. 1999). This finding suggested the presence of an additional presynaptic receptor. In α_{2A}-KO tissues, any correlation of antagonist pK_d values with those of α_{2D}-pharmacology was lost and the remaining receptors displayed α_{2B}- or α_{2C}-pharmacology (Trendelenburg et al. 1999). By crossing mouse lines deficient in single α_2-receptors, double transgenic mice could be generated which lacked both α_{2A}- and α_{2C}-receptors (α_{2AC}-KO) (Hein et al. 1999). Using these α_{2AC}-KO mice, the second presynaptic receptor could be identified as the α_{2C}- receptor, as no agonist effect remained in tissues from mice lacking α_{2A}- and α_{2C}-receptors (Fig. 4) (Hein et al. 1999). Both, α_{2A}- and α_{2C}-receptors operated as presynaptic inhibitory receptors in a wide range of tissues investigated so far, including central adrenergic neurons, e.g. brain cortex and hippocampus, and peripheral sympathetic neurons in heart atria and vas deferens (Hein et al. 1999; Trendelenburg et al. 1999). Pharmacological studies in human atria suggested that indeed α_{2C}-receptors are involved in presynaptic autoregulation of noradrenaline from sympathetic nerves (Rump et al. 1995). In the brain, subtle changes have been observed in dopamine and serotonin metabolism in transgenic mice with altered expression of the α_{2C}-receptor (Sallinen et al. 1997). Lack of α_{2C}-receptor expression (α_{2C}-KO) slightly decreased the rate of monoamine turnover in the brain and overexpression of α_{2C}-receptors increased dopamine stores and metabolism (Sallinen et al. 1997).

Several lines of evidence suggest that α_{2A}- and α_{2C}-receptors operate also in wild-type mice as integral parts of the presynaptic feedback loop. Experiments on peripheral tissues, e.g. heart atria, demonstrated that the α_{2C}-subtype mediates autoinhibition by low concentrations of noradrenaline in wild-type mice, whereas the potency of noradrenaline at the α_{2A}-subtype was lower (Fig. 4) (Hein et al. 1999). This potency difference of noradrenaline for the α_{2}-receptors correlated with the affinity difference of noradrenaline for the α_{2A}- and α_{2C}-subtypes, respectively. Furthermore, there is no evidence so far that the expression of the remaining α_{2}-receptor subtypes was altered in mice carrying deletions in α_{2A}- or α_{2C}-receptor genes (Link et al. 1995; Altman et al. 1999).

Presynaptic α_{2A}- and α_{2C}-receptors can be distinguished functionally and may thus serve independent presynaptic functions. In mouse atria, the α_{2A}-subtype inhibited noradrenaline release at high stimulation frequencies whereas the α_{2C}-receptors operated at lower levels of sympathetic nerve stimulation (Hein et al. 1999). Moreover, inhibition of noradrenaline release mediated by the α_{2A}-subtype occurred much faster than inhibition by the α_{2C}-receptor. These findings indicate that two presynaptic receptors in the inhibitory feedback loop of transmitter release may differentially regulate synaptic transmission.

In mice carrying the mutated α_{2A}-D79N receptor gene, presynaptic function in the vas deferens did not differ from wild-type control mice (Altman et al. 1999). This finding indicates (1) that K^+ channel coupling is not required for presynaptic function of α_{2A}-receptors in sympathetic nerves and (2) that there is a high number of spare receptors at the presynaptic site, as expression of the α_{2A}-D79N receptor was found to be reduced to 20% of the wild-type level (MacMillan et al. 1996). Thus, Ca^{2+} channels in presynaptic nerve terminals are essential for neurotransmitter release, and current research has provided evidence for the involvement of a multitude of Ca^{2+} channel subtypes (Reuter 1996).

4 Cardiovascular Function of α_2-Receptors

4.1 Hemodynamic Effects of α_2-Agonists

α_2-Adrenergic receptors in the rostral ventrolateral medulla of the brain respond to noradrenaline and adrenaline to decrease sympathetic outflow and reduce arterial pressure and heart rate (Ruffolo et al. 1993). This hypotensive effect of α_2-agonists has been the rationale for the therapeutic use of clonidine in the treatment of hypertension (Ruffolo et al. 1993). However, especially after rapid intravenous injections, clonidine may initially cause a transient hypertensive response mediated by α_2-receptor-elicited contraction of the peripheral vasculature. Using α_2-subtype-specific knockout mice, these two opposing hemodynamic effects of an α_2-agonist could be identified as being mediated by the α_{2A}- (hypotension) and the α_{2B}-receptor (hypertension). In conscious, unrestrained wild-type mice, infusions of the α_2-agonist brimonidine (UK14,304) or dexmedetomidine into the carotid artery resulted in a transient pressor response followed by an extended hypotensive response. The hypotensive response was essentially absent in α_{2A}-KO and in α_{2A}-D79N mice (MacMillan et al. 1996; Altman et al. 1999). However, the rapid initial hypertensive response to α_2-agonist infusion was abolished in α_{2B}-KO mice and the hypotensive response occurred immediately and was significantly greater in α_{2B}-KOs than in wild-type mice (Link et al. 1996). These results demonstrate that stimulation of α_{2B}-receptors counteracts the therapeutic antihypertensive effect of drugs acting at central α_{2A}-receptors in the central nervous system (Link et al. 1996). In addition, the α_{2A}-receptor may contribute to the vasoconstriction in some vascular beds, as the hypertensive response to α_2-agonists in α_{2A}-D79N mice was absent after femoral administration but unchanged after carotid administration after carotid injection of α_2-agonists (MacMillan et al. 1996).

4.2 Baseline Hemodynamics of α_2-Receptor-Deficient Mice

At baseline, α_{2B}-KO, α_{2C}-KO, and α_{2A}-D79N mice had similar heart rate and blood pressure as compared with wild-type mice (Link et al. 1996; MacMillan et al. 1996). However, α_{2A}-KO mice were tachycardic at rest and showed a significant increase in arterial pressure (Makaritsis et al. 1999). The increased heart rate in α_{2A}-KO was due to an increase in sympathetic activity, as infusion of the β-receptor antagonist propranolol could completely abolish the chronotropic effect of the gene deletion (Altman et al. 1999). As expected from the presynaptic location of the α_{2A}- and α_{2C}-receptor subtypes, plasma noradrenaline levels were slightly elevated in α_{2A}-KO mice and were 3-fold above the control value in α_{2AC}-KO mice lacking both presynaptic receptors (Hein et al. 1999). Increased noradrenaline turnover was also observed in mice expressing the α_{2A}-D79N receptor (Lakhlani et al. 1997).

4.3 Long-Term Effects of α_2-Receptor Deletion on the Cardiovascular System

The long-term physiological consequences of altered sympathetic transmitter release were studied in mice lacking α_{2A}- and α_{2C}-receptors. The heart is very sensitive to chronic elevations of catecholamines, and abnormal activity of the sympathetic nervous system has been implicated in the pathogenesis of heart failure (Cohn et al. 1984; Francis et al. 1990; Packer 1992; Lechat et al. 1998). In α_{2AC}-KO mice, left ventricular maximal contractility decreased to 70% of the wild-type value, while mice lacking single α_2-receptor subtypes did not show an impairment of cardiac function (Hein et al. 1999). In addition, cardiac hypertrophy developed in α_{2AC}-KO mice. These findings indicate that the heart is exposed to significantly higher levels of catecholamines in α_{2AC}-KO mice than in either α_{2A}-KO or α_{2C}-KO mice.

The role of the α_{2B}-receptor in cardiovascular regulation was further emphazised by results from experiments using a hypertension model (Makaritsis et al. 1999). The increase in blood pressure after subtotal nephrectomy and replacement of the drinking water with 1% saline was significantly smaller in α_{2B}-KO than in wild-type or α_{2C}-KO mice (Makaritsis et al. 1999). Thus, α_{2B}-receptors are necessary to raise blood pressure in response to salt-loading. It is unclear whether this process is of central origin (inability to increase sympathetic outflow), vascular origin (inability to vasoconstrict) or renal origin (inability to retain excess salt and fluid) (Makaritsis et al. 1999).

4.4 Imidazoline Receptors

In 1984, Bousquet et al. suggested that a characteristic cerebral effect of clonidine-like imidazoline derivatives, hypotension, might be mediated not by α_2-adrenergic receptors but by separate "imidazoline-preferring" receptors. Imidazoline receptors are now being discussed as potential sites of drug action in many tissues and are thought to comprise several distinct types, of which two have been termed I_1 and I_2 (for review, see Molderings 1997). There is controversy concerning whether agents such as clonidine, which contain an imidazole moiety, elicit their hypotensive effects by interacting with α_2-receptors or with a separate imidazoline receptor population (Bousquet et al. 1992). To address this question, the hypotensive effect of several imidazoline ligands was tested in α_{2A}-D79N mice (Zhu et al. 1999). In wild-type mice, rilmenidine, moxonidine and clonidine decreased blood pressure and heart rate. In α_{2A}-D79N mice, hypotensive responses to rilmenidine and moxonidine were completely absent. After clonidine infusion, hypotension was absent in α_{2A}-D79N animals. In contrast, dose-dependent hypertension and bradycardia were observed. Thus, there was no evidence for imidazoline receptor-mediated hemodynamic effects in mice carrying a mutated α_{2A}-receptor gene (Zhu et al. 1999).

5 Central Nervous System Function of α_2-Receptor Subtypes

5.1 Sedation

α_2-adrenergic agonists are used in the anaesthetic management of the surgical patient for their sedative and hypnotic properties. The sedative effects of α_2-agonists was investigated in α_{2A}-D79N mice and in mice with altered expression of the α_{2C}-receptor (knockout and overexpression) (Hunter et al. 1997; Lakhlani et al. 1997). α_{2A}-D79N mice showed no sedative response to the α_2-agonist, dexmedetomidine, indicating that the α_{2A}-subtype is responsible for the clinically used hypnotic effect of α_2-agonists (Lakhlani et al. 1997). In mice lacking α_{2B}- or α_{2C}-receptors, the sedative response to dexmedetomidine was unaltered when compared with wild-type mice (Hunter et al. 1997; Sallinen et al. 1997). Transgenic overexpression of the α_{2C}-receptor did not affect cortical EEG delta amplitudes at baseline and after α_2-receptor activation (Björklund et al. 1998).

A clinically useful action of α_2-agonists is their ability to reduce the requirements for other anaesthetic agents during anaesthesia. In control mice, non-sedative doses of dexmedetomidine reduced the concentrations of the volatile anaesthetic, halothane, to induce anaesthesia by 30% (Lakhlani et al. 1997). This anaesthetic-sparing effect of α_2-agonists was completely abolished in α_{2A}-D79N mice.

Several lines of evidence suggest that the locus coeruleus is the site of the α_{2A}-agonist-mediated sedative response. The α_{2A}-subtype is abundantly expressed in locus coeruleus neurons (Wang et al. 1996). Administration of antisense oligonucleotides for the α_{2A}-receptor subtype into the locus coeruleus of rats attenuated the hypnotic effect of dexmedetomidine reversibly (Mizobe et al. 1996). In the locus coeruleus of wild-type mice, α_2-agonists suppressed the spontaneous firing rate of neurons but did not alter spontaneous activity or membrane potential in neurons from α_{2A}-D79N mice (Lakhlani et al. 1997). In addition, α_2-receptor activation reduces Ca^{2+} channel currents in numerous neuronal preparations. The inhibitory effect of the α_2-agonists, clonidine and dexmedetomidine on voltage-gated Ca^{2+} channels in locus coeruleus or superior cervical ganglion cells was significantly blunted, but not abolished in α_{2A}-D79N mice (Lakhlani et al. 1997). In α_{2A}-D79N neurons, the inhibitory effect of clonidine on Ca^{2+} currents was reduced to 30% of the response recorded in cells from wild-type mice and this effect was sensitive to the α-receptor antagonist prazosin, indicating that α_{2B}- or α_{2C}-receptors are involved in this response.

5.2 Analgesia

Analgesia is a clinically important use of α_2-receptor agonists. α_2-Agonists mediate analgesia (Yaksh 1985) and they interact synergistically with opioids (Sullivan et al. 1987; Wilcox et al. 1987; Drasner and Fields 1988; Ossipov et al. 1989; Monasky et al. 1990). Pharmacological studies have suggested that activation of α_{2A}-receptors mediates the α_2-agonist-induced analgesia (Millan 1992; Millan et al. 1994), others have suggested that the site of action may be α_{2A}- or non-α_{2A}-receptors dependent on the agonist used (Takano and Yaksh 1992). Subtype-selective antisera have localized the α_{2A}-subtype in the rat spinal cord to terminals of capsaicin-sensitive, substance P-containing primary afferent fibers (Stone et al. 1998). The α_{2C}-subtype was found to be expressed in a subset of spinal interneurons. In situ hybridization studies have localized mRNA for α_{2A}- and α_{2C}-subtypes in dorsal root ganglion neurons (Nicholas et al. 1993), thus one or both subtypes may mediate spinal analgesia at a presynaptic site on primary afferent fibers.

In comparison to control mice, dexmedetomidine was completely ineffective as an antinociceptive agent in the tail immersion test in the α_{2A}-D79N transgenic mice (Hunter et al. 1997a; Hunter et al. 1997b). In α_{2A}-D79N mice, intrathecal administration of brimonidine (UK14,304) had no analgesic effect in the tail-flick test, whereas the analgesic potency of morphine was not altered in these mice (Stone et al. 1997). The α_{2A}-D79N mutation also decreased α_2-agonist-mediated spinal analgesia and blocked the synergy seen in wild-type mice with δ-opioid or μ-opioid agonists in the substance P behavioral test (Stone et al. 1997). However, some α_2-agonist effect remained in the substance P test in α_{2A}-D79N mice which could be attributed to residual adenylyl cyclase coupling of the α_{2A}-D79N receptor or to another α_2-subtype. Combinations of α_{2A}-agonists and μ-opioid agonists may prove useful in maximizing the analgesic efficacy of opioids while decreasing total dose requirements.

Supraspinal opioid receptors and spinal α_2-receptors are involved in the analgesic mechanism for nitrous oxide (N_2O). It has been suggested that activation of opioid receptors in the periaqueductal gray activates descending noradrenergic pathways which release noradrenaline onto α_2-receptors in the dorsal horn of the spinal cord (Zhang et al. 1999). After exposure of rats to N_2O a fourfold increase in noradrenaline release could be detected in the dorsal horn of the spinal cord (Zhang et al. 1999). N_2O produced antinociception in the tail flick test in wild-type and in α_{2A}-D79N mice, although the response was less pronounced in α_{2A}-D79N mice (Guo et al. 1999). The antinociceptive response to N_2O in α_{2A}-D79N mice could be antagonized by opioid receptor antagonists and by prazosin, which blocks α_{2B}- and α_{2C}-receptors. Adrenergic agonists have been shown to inhibit neurotransmitter release from spinal cord preparations by a prazosin-sensitive receptor, suggesting a role for the α_{2B} or α_{2C} subtypes (Ono et al. 1991). Thus, the analgesic effect of α_2-agonists seems to be mediated by the α_{2A}-receptor subtype. In addition, α_{2B}- and/or α_{2C}-receptors may be involved in the antinociceptive effect of nitrous oxide.

5.3 Inhibition of Epileptic Seizures

Noradrenaline is unique among the monoamine transmitters in that it exerts powerful antiepileptogenic actions (McNamara et al. 1987) which are mediated by the α_2-receptor (Gellman et al. 1987). Mice carrying a mutated α_{2A}-receptor (α_{2A}-D79N) showed marked enhancement of epileptogenesis in a mouse kindling model, and the proepileptogenic actions of the α_2-antagonist idazoxan were abolished (Janumpalli et al. 1998). These data suggest that the α_{2A}-subtype is the only α_2-receptor involved in modulating seizure threshhold.

5.4 Hypothermia

α_2-Agonists lower body temperature dose-dependently (Hunter et al. 1997b; Sallinen et al. 1997). Dexmedetomidine showed a hypothermic effect in mice lacking α_{2B}- or α_{2C}-receptors but failed to decrease body temperature in α_{2A}-D79N mice (Hunter et al. 1997). Contrary to these results, the hypothermic effect of dexmedetomidine was slightly blunted in α_{2C}-KO mice (Sallinen et al. 1997) and it had no effect on body temperature in α_{2C}-KO mice at a low dose that produced significant hypothermia in wild-type mice (Hunter et al. 1997b). Thus, two α_2-receptor subtypes, α_{2A} and α_{2C}, may be involved in the regulation of body temperature.

5.5 Behavioural Functions

α_2-Adrenergic receptors mediate many physiological functions and pharmacological effects in the central nervous system, mainly by inhibiting neuronal firing and release of noradrenaline and other neurotransmitters. Locus coeruleus noradrenaline neurons send noradrenergic fibers into different forebrain structures and modulate different cognitive functions, such as attention, arousal, and planning (Crow 1968; Arnsten and Goldman-Rakic 1985; Arnsten and Leslie 1991; Riekkinen et al. 1992; Arnsten et al. 1996; Coull et al. 1996). A variety of behavioural paradigms were tested in mice lacking or overexpressing α_{2C}-adrenergic receptor, but no data have been obtained for α_{2A}- or α_{2B}-receptor-deficient mice.

Activation of α_2-receptors resulted in locomotor inhibition. The α_2-agonist dexmedetomidine did not alter spontaneous motor activity or diurnal rhythm of motor activity of α_{2C}-KO or α_{2C}-overexpressing mice (Sallinen et al. 1997). Thus, the α_{2C}-subtype does not seem to be involved in the effect α_2-agonists on locomotor behaviour. However, D-amphetamine stimulated locomotor activity to a greater extent in α_{2C}-KO mice than in wild-type mice (Sallinen et al. 1998). The behavioural serotonin syndrome and head twitches to injection of the serotonin precursor 5-hydroxytryptophan were inhibited by α_2-agonists with similar magnitude in wild-type and α_{2C}-KO mice, suggesting that the α_{2A}-subtype rather than the α_{2C}-receptor may be involved in α_2-mediated inhibition of the serotonin syndrome (Sallinen et al. 1998).

Experimental data indicate that antagonists selective for the α_{2C}-subtype and agonists devoid of any α_{2C}-receptor affinity can modulate cognition more favourably than subtype-nonselective drugs. Mice overexpressing α_{2C}-receptors were impaired in spatial or nonspatial water maze tests, and an α_2-antagonist fully reversed the water maze escape defect in α_{2C}-receptor overexpressing mice (Björklund et al. 1998; Björklund et al. 1999; Björklund et al. 2000). The α_2-agonist dexmedetomidine increased swimming distance more effectively in wild-type mice than in α_{2C}-KO mice (Björklund et al. 1998). These results suggest that α_{2C}-receptors can modulate navigation to a hidden or visible escape platform. Activation of the α_{2C}-subtype disrupts execution of spatial and non-spatial search patterns (Björklund et al. 1999).

Altered startle reactivity and attenuation of the inhibition of the startle reflex by an acoustic prepulse has been observed in psychiatric patients, e.g. in schizophrenia (Braff et al. 1978). Disrupted prepulse inhibition in rats can be normalized by antipsychotics and this paradigm is being used as an animal model for drug development. Interestingly, α_{2C}-KO mice had enhanced startle responses, diminished prepulse inhibition, and shortened attack latency in the isolation-aggression test (Sallinen et al. 1998). The opposite effect was observed in mice overexpressing the α_{2C}-receptor. Thus drugs acting via the α_{2C}-receptor might have therapeutic value in disorders associated with enhanced startle responses and sensorimotor gating deficits, such as schizophrenia, attention deficit disorder, post-traumatic stress disorder, and drug withdrawal. Activation of α_{2C}-receptors reduces hyperreactivity and impulsivity of mice, indicating that the α_{2C}-subtype has an inhibitory role on reactivity of the central nervous system. It is tempting to speculate that the therapeutic benefit and clinical acceptance of clonidine in neuropsychiatric disorders might have been restricted by the adverse effects of hypotension and sedation, which seem to be mediated solely by α_{2A}-receptors. However, recent studies in mouse behavioural models suggest that the α_{2A}-receptor has a protective role in some forms of depression and anxiety, and that this subtype mediates the antidepressant effects of imipramine (Schramm, McDonald, Limbird, personal communication). Thus, α_{2A}- and α_{2C}-receptors may complement each other to integrate central nervous system function.

6 Conclusion

Gene targeting in mice represents a unique approach to delineate the physiological functions of individual α_2-adrenergic receptor subtypes in vivo (Fig. 5). In the cardiovascular system, stimulation of α_{2B}-receptors in vascular smooth muscle cells produces hypertension and counteracts the clinically beneficial hypotensive effect of stimulating α_{2A}-receptors in the central nervous system. In addition to the hypotensive action of α_2-agonists, their hypnotic, antiepileptogenic, and analgesic effects are mediated via the α_{2A}-adrenergic subtype. However, some evidence accumulates that the α_{2C}-receptor may be an important target for fine tuning of neurotransmitter release in the central and peripheral nervous system. In addition, several aspects of cognitive function which are modulated by α_2-agonists are mediated via the α_{2C}-subtype. However, further studies are required, as not all of the clinically relevant functions of α_2-receptors have been systematically investigated in all three lines of α_2-receptor-deficient mice simultaneously. Rather, many assignments of α_{2A}-receptor functions have been made by using mice carrying a point mutation (D79N) in the α_{2A}-receptor gene, which partially uncouples this receptor mutant from intracellular signalling pathways and dramatically reduces receptor expression in vivo (MacMillan et al. 1996; Lakhlani et al. 1997). The α_{2A}-D79N transgenic mouse provides important in vivo insight into the physiological significance of individual signalling pathways of an individual receptor. However, caution should be

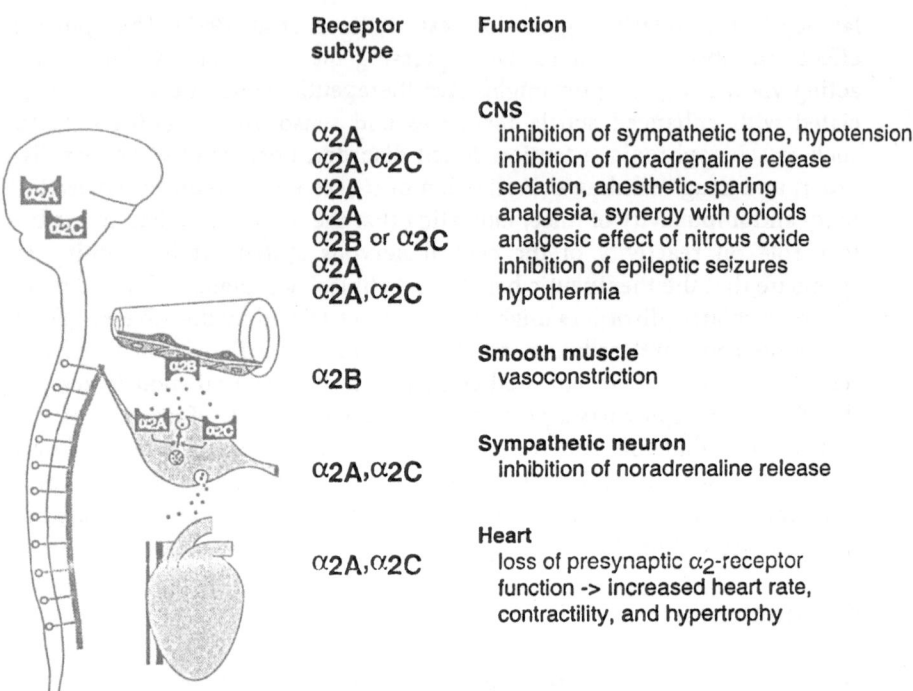

Receptor subtype	Function
	CNS
α2A	inhibition of sympathetic tone, hypotension
α2A,α2C	inhibition of noradrenaline release
α2A	sedation, anesthetic-sparing
α2A	analgesia, synergy with opioids
α2B or α2C	analgesic effect of nitrous oxide
α2A	inhibition of epileptic seizures
α2A,α2C	hypothermia
	Smooth muscle
α2B	vasoconstriction
	Sympathetic neuron
α2A,α2C	inhibition of noradrenaline release
	Heart
α2A,α2C	loss of presynaptic α2-receptor function -> increased heart rate, contractility, and hypertrophy

Fig. 5. In vivo functions of α_2-adrenergic receptor subtypes. This overview assigns physiological effects of α_2-receptor stimulation to individual α_2-receptor subtypes based on experiments performed in transgenic mice carrying deletions in α_2-receptor gene (see text for references)

used when interpreting data obtained with this mouse model as a "functional knockout". In addition, double knockout mice lacking two of the three α_2-receptor subtypes will be important tools to assign α_2-receptor functions more precisely to individual subtypes. The combined deletion of α_{2A}- and α_{2C}-receptor (α_{2AC}-KO) highlights the importance of combining transgenic lines to define the function of the presynaptic α_2-receptor feedback loop for neurotransmitter release. With these tools, in vivo gene targeting will be further exploited to identify the pharmacological significance of specific receptor subtypes in vivo and to guide pharmaceutical development of novel subtype-selective drugs.

Acknowledgements

The authors work has been supported by the Deutsche Forschungsgemeinschaft SFB355 and SFB 487. The author is indebted to Martin J. Lohse, Würzburg, for critical reading of the manuscript.

References

Alblas J, van Corven EJ, Hordijk PL, Milligan G, Moolenaar WH (1993) G_i-mediated activation of the p21ras-mitogen-activated protein kinase pathway by α_2-adrenergic receptors expressed in fibroblasts. J Biol Chem 268: 22235-22238

Altman JD, Trendelenburg AU, MacMillan L, Bernstein D, Limbird L, Starke K, Kobilka BK, Hein L (1999) Abnormal regulation of the sympathetic nervous system in α_{2A}-adrenergic receptor knockout mice. Mol Pharmacol 56: 154-161

Anderson NG, Milligan G (1994) Regulation of p42 and p44 MAP kinase isoforms in Rat-1 fibroblasts stably transfected with α_2C10 adrenoreceptors. Biochem Biophys Res Commun 200: 1529-1535

Aoki C, Go CG, Venkatesan C, Kurose H (1994) Perikaryal and synaptic localization of α_{2A}-adrenergic receptor-like immunoreactivity. Brain Res 650: 181-204

Arnsten AFT, Goldman-Rakic PS (1985) α_2 Adrenergic mechanism in prefrontal cortex associated with cognitive decline in aged nonhuman primates. Science 230: 1273-1376

Arnsten AFT, Leslie FM (1991) Behavioural and receptor binding analysis of the α_2-agonist, UK-14304 (5 bromo-6 (2-imidazoline-2-yl amino) quinoxaline); evidence for cognitive enhancement at an α_2-adrenoceptor subtype. Neuropharmacology 30: 1279-1289

Arnsten AFT, Steere JC, Hunt RD (1996) The contribution of α_2-noradrenergic mechanisms to prefrontal cortical cognitive function. Arch Gen Psychiatry 53: 448-455

Björklund M, Sirviö J, Pouliväli J, Sallinen J, Jäkälä P, Scheinin M, Kobilka BK, Riekkinen PJr (1998) α_{2C}-Adrenoceptor-overexpressing mice are impaired in executing nonspatial and spatial escape strategies. Mol Pharmacol 54: 569-576

Björklund M, Sirviö J, Riekkinen PJ, Sallinen J, Scheinin M, Riekkinen PJ (2000) Overexpression of α_{2C}-adrenoceptors impairs water maze navigation. Neuroscience 95: 481-487

Björklund M, Sirviö J, Sallinen J, Scheinin M, Kobilka BK, Riekkinen PJ (1999) α_{2C}-adrenoceptor overexpression disrupts execution of spatial and non-spatial search patterns. Neuroscience 88: 1187-1198

Bousquet P, Feldman J, Tibirica E, Bricca G, Greney H, Dontenwill M, Stutzmann J, Belcourt A (1992) Imidazoline receptors. A new concept in central regulation of the arterial blood pressure. Am J Hypertens 5: 47S-50S

Braff D, Grillon C, Gallaway E, Geyer M, Bali L (1978) Prestimulus effects on human startle reflex in normal and schizophrenics. Psychophysiology 15: 339-343

Bylund DB, Eikenberg DC, Hieble JP, Langer SZ, Lefkowitz RJ, Minneman KP, Molinoff PB, Ruffolo RR, Trendelenburg U (1994) International Union of Pharmacology nomenclature of adrenoceptors. Pharmacol Rev 46: 121-136

Chung FZ, Wang CD, Potter PC, Venter JC, Fraser CM (1988) Site-directed mutagenesis and continuous expression of human β-adrenergic receptors.

Identification of a conserved aspartate residue involved in agonist binding and receptor activation. J Biol Chem 263: 4052-4055

Cohn JF, Levine TB, Olivari MT, Garberg V, Lura D, Francis GS, Simon AB, Rector T (1984) Plasma norepinephrine as a guide to prognosis in patients with chronic congestive heart failure. N Engl J Med 311: 819-823

Cotecchia S, Kobilka BK, Daniel KW, Nolan RD, Lapetina EY, Caron MG, Lefkowitz RJ, Regan JW (1990) Multiple second messenger pathways of α-adrenergic receptor subtypes expressed in eukaryotic cells. J Biol Chem 265: 63-69

Coull JT, Sahakian BJ, Hodges JR (1996) The α_2 antagonist idazoxan remediates certain attentional and executive dusfunction in patients with dementia of frontal type. Phychopharmacology 123: 239-249

Crow TJ (1968) Cortical synapses and reinforcement. Nature 219: 736-737

Daunt DA, Hurt C, Hein L, Kallio J, Feng F, Kobilka BK (1997) Subtype-specific intracellular trafficking of α_2-adrenergic receptors. Mol Pharmacol 51: 711-720

DeGraff JL, Gagnon AW, Benovic JL, Orsini MJ (1999) Role of arrestins in endocytosis and signalling of α_2-adrenergic receptor subtypes. J Biol Chem 274, 11253-11259

Delmas P, Abogadie FC, Milligan G, Buckley NJ, Brown DA (1999) $\beta\gamma$ dimers derived from G_o and G_i proteins contribute different components of adrenergic inhibiton of Ca^{2+} channels in rat sympathetic neurons. J Physiol 518: 23-36

Diverse-Pierluissi M, Inglese J, Stoffel RH, Lefkowitz RJ, Dunlap K (1996) G protein-coupled receptor kinase mediates desensitization of norepinephrine-induced Ca^{2+} channel inhibition. Neuron 16: 579-585

Docherty JR (1998) Subtypes of functional α_1- and α_2-adrenoceptors. Eur J Pharmacol 361: 1-15

Drasner K, Fields HL (1988) Synergy between the antinocipeptive effects of intrathecal clonidine and systemic morphine in the rat. Pain 32: 309-312

Dzimiri N (1999) Regulation of β-adrenoceptor signaling in cardiac function and disease. Pharmacol Rev 51: 465-501

Eason MG, Jacinto MT, Liggett SB (1994) Contribution of ligand structure to activation of α_2-adrenergic receptor subtype coupling to G_s. Molec Pharmacol 45: 696-702

Eason MG, Kurose H, Holt BD, Raymond JR, Liggett SB (1992) Simultaneous coupling of α_2-adrenergic receptors to two G proteins with opposing effects. Subtype-selective coupling of α_2C10, α_2C4, and α_2C2 adrenergic receptors to G_i and G_s. J Biol Chem 267: 15795-15801

Eason MG, Liggett SB (1992) Subtype-selective desensitization of α_2-adrenergic receptors. Different mechanisms control short and long term agonist-promoted desensitization of α_2C10, α_2C4, and α_2C2. J Biol Chem 267: 25473-25479

Eason MG, Liggett SB (1995) Identification of a G_s coupling domain in the amino terminus of the third intracellular loop of the α_{2A}-adrenergic receptor. Evidence for distinct structural determinants that confer G_s versus G_i coupling. J Biol Chem 270: 24753-24760

Francis GS, Benedict C, Johnstone DE, Kirlin PC, Nicklas J, Liang CS, Kubo SH, Rudin-Toretsky E, Yusuf S (1990) Comparison of neuroendocrine activation in patients with left ventricular dysfunction with and without congestive heart failure: a substudy of the studies of left ventricular dysfunction (SOLVD). Circulation 82: 1724-1729

Fraser CM, Wang CD, Robinson DA, Gocayne JD, Venter JC (1989) Site-directed mutagenesis of m1 muscarinic acetylcholine receptors: conserved aspartic acids play important roles in receptor function. Mol Pharmacol 36: 840-847

Gellman RL, Kallianos JA, McNamara JO (1987) α_2 receptors mediate endogenous noradrenergic suppression of kindling development. J Pharmacol Exp Ther 241: 891-898

Guo TZ, Davies MF, Kingery WF, Patterson AJ, Limbird LE, Maze M (1999) Nitrous oxide produces antinociceptive response via α_{2B} and/or α_{2C} adrenoceptor subtypes in mice. Anesthesiology 90: 470-476

Guyer CA, Horstman DA, Wilson AL, Clark JD, Cragoe EJ Jr, Limbird LE (1990) Cloning, sequencing, and expression of the gene encoding the porcine α_2-adrenergic receptor. Allosteric modulation by Na^+, H^+, and amiloride analogs. J Biol Chem 265: 17307-17317

Hein L, Altman JD, Kobilka BK (1999) Two functionally distinct α_2-adrenergic receptors regulate sympathetic neurotransmission. Nature 402: 181-184

Hunter JC, Fontana DJ, Hedley LR, Jasper JR, Kassotakis L, Lewis R, Eglen RM (1997a) The relative contribution of α_2-adrenoceptor subtypes to the antinociceptive action of dexmedetomidine and clonidine in rodent models of acute and chronic pain. Br J Pharmacol 120: 229P

Hunter JC, Fontana DJ, Hedley LR, Jasper JR, Lewis R, Link RE, Secchi R, Sutton J, Eglen RM (1997b) Assessment of the role of α_2-adrenoceptor subtypes in the antinociceptive, sedative and hypothermic action of dexmedetomidine in transgenic mice. Br J Pharmacol 122: 1339-1344

Janumpalli S, Butler LS, MacMillan LB, Limbird LE, McNamara JO (1998) A point mutation (D79N) of the α_{2A} adrenergic receptor abolishes the antiepileptogenic action of endogenous norepinephrine. J Neurosci 18: 2004-2008

Jeong SW, Ikeda SR (2000) Effect of G protein heterotrimer composition on coupling of neurotransmitter receptors to N-type Ca^{2+} channel modulation in sympathetic neurons. Proc Natl Acad Sci USA 97: 907-912

Jewell-Motz EA, Liggett SB (1996) G protein-coupled receptor kinase specificity for phosphorylation and desensitization of α_2-adrenergic receptor subtypes. J Biol Chem 271: 18082-18087

Kable JW, Murrin LC, Bylund DB (2000) In vivo gene modification elucidates subtype-specific fucntions of α_2-adrenergic receptors. J Pharmacol Exp Ther 293, 1-7

Koch WJ, Hawes BE, Allen LF, Lefkowitz RJ (1994) Direct evidence that G_i-coupled receptor stimulation of mitogen-activated protein kinase is mediated by G $\beta\gamma$ activation of p21ras. Proc Natl Acad Sci U S A 91: 12706-12710

Kurose H, Lefkowitz RJ (1994) Differential desensitization and phosphorylation of three cloned and transfected α_2-adrenergic receptor subtypes. J Biol Chem 269: 10093-10099

Lakhlani PP, MacMillan LB, Guo TZ, McCool BA, Lovinger DM, Maze M, Limbird LE (1997) Substitution of a mutant α_{2A}-adrenergic receptor via "hit and run" gene targeting reveals the role of this subtype in sedative, analgesic, and anesthetic-sparing responses in vivo. Proc Natl Acad Sci U S A 94: 9950-9955

Langer SZ (1974) Presynaptic regulation of catecholamine release. Biochem Pharmacol 23: 1793-1800

Langer SZ (1997) 25 years since the discovery of presynaptic receptors: present knowledge and future perspectives. Trends Pharmacol Sci 18: 95-99

Lechat P, Packer M, Chalon S, Cucherat M, Arab T, Boissel JP (1998) Clinical effects of β-adrenergic blockade in chronic heart failure: a meta-analysis of double-blind, placebo-controlled, randomized trials. Circulation 98: 1184-1191

Liggett SB (1999) Molecular and genetic basis of β_2-adrenergic receptor function. J Allergy Clin Immunol 104: S42-S46

Link RE, Daunt D, Barsh G, Chruscinski A, Kobilka BK (1992) Cloning of two mouse genes encoding α_2-adrenergic receptor subtypes and identification of a single amino acid in the mouse α_2-C10 homolog responsible for an interspecies variation in antagonist binding. Molec Pharmacol 42: 16-27

Link RE, Desai K, Hein L, Stevens ME, Chruscinski A, Bernstein D, Barsh GS, Kobilka BK (1996) Cardiovascular regulation in mice lacking α_2-adrenergic receptor subtypes b and c. Science 273: 803-805

Link RE, Stevens MS, Kulatunga M, Scheinin M, Barsh GS, Kobilka BK (1995) Targeted inactivation of the gene encoding the mouse α_{2C}-adrenoceptor homolog. Mol Pharmacol 48: 48-55

Lowell BB (1998) Using gene knockout and transgenic techniques to study the physiology and pharmacology of β_3-adrenergic receptors. Endocr J 45: S9-S13

MacDonald E, Kobilka BK, Scheinin M (1997) Gene targeting: homing in on α_2-adrenoceptor-subtype function. Trends Pharmacol Sci 18: 211-219

MacDonald E, Scheinin M (1995) Distribution and pharmacology of α_2-adrenoceptors in the central nervous system. J Physiol Pharmacol 46: 241-258

MacMillan LB, Hein L, Smith MS, Piascik MT, Limbird LE (1996) Central hypotensive effects of the α_{2A}-adrenergic receptor subtype. Science 273: 801-803

MacMillan LB, Lakhlani PP, Hein L, Piascik M, Guo TZ, Lovinger D, Maze M, Limbird LE (1998) In vivo mutation of the α_{2A}-adrenergic receptor by homologous recombination reveals the role of this receptor subtype in multiple physiological processes. Adv Pharmacol 42: 493-496

MacNulty EE, McClue SJ, Carr IC, Jess T, Wakelam MJ, Milligan G (1992) α_2-C10 adrenergic receptors expressed in rat 1 fibroblasts can regulate both adenylylcyclase and phospholipase D-mediated hydrolysis of phosphatidylcholine by interacting with pertussis toxin-sensitive guanine nucleotide-binding proteins. J Biol Chem 267: 2149-2156

Makaritsis KP, Handy DE, Johns C, Kobilka BK, Gavras I, Gavras H (1999) Role of the α_{2B}-adrenergic receptor in the development of salt-induced hypertension. Hypertension 33: 14-17

Makaritsis KP, Johns C, Gavras I, Altman JD, Handy DE, Bresnahan MR, Gavras H (1999) Sympathoinhibitory function of the α_{2A}-adrenergic receptor subtype. Hypertension 34: 403-407

McNamara JO, Bonhaus DW, Crain BJ, Gellman RJ, Shin C (1987) Biochemical and pharmacological studies of neurotransmitters in the kindling model. In: Neurotransmitters and epilepsy. Jobe PC and Laird HE (Eds) Clifton, NJ, Humana: 115-160

Millan MJ (1992) Evidence that an α_{2A}-adrenoceptor subtype mediates antinociception in mice. Eur J Pharmacol 215: 355-356

Millan MJ, Bervoets K, Rivet JM, Widdowson P, Renourd A, LeMarouille-Girardon S, Gobert A (1994). Multiple α_2-adrenergic receptor subtypes. II. Evidence for a role of rat α_{2A} adrenergic receptors in the control of nociception: motor behavior and hippocampal synthesis of noradrenaline. J Pharmacol Exp Ther 270: 958-972

Mizobe T, Maghsoudi K, Sitwala K, Tianzhi G, Ou J, Maze M (1996) Antisense technology reveals the α_{2A} adrenoceptor subtype to be the subtype mediating the hypnotic response to the highly selective agonist, dexmedetomidine, in the locus coeruleus of the rat. J Clin Invest 98: 1076-1080

Molderings GJ (1997) Imidazoline receptors: basic knowledge, recent advances and future prospects for therapy and diagnosis. Drugs Future 22: 757-772

Monasky MS, Zinsmeister AR, Stevens CW, Yaksh TL (1990) Interaction of intrathecal morphine and ST-91 on antinociception in the rat: dose-response analysis, antagonism and clearance. J Pharm Exp Ther 254: 383-392

Nicholas AP, Hökfelt T, Pieribone VA (1996) The distribtion and significance of CNS adrenoceptors examined with in situ hybridization. Trends Pharmacol Sci 17: 245-255

Nicholas AP, Pieribone V, Hökfelt T (1993) Distributions of mRNAs for α_2 adrenergic receptor subtypes in rat brain: an in situ hybridization study. J Comp Neurol 328: 575-594

Olli-Lähdesmäki T, Kallio J, Scheinin J (1999) Receptor subtype-induced targeting and subtype-specific internalization of human α_2-adrenoceptors in PC12 cells. J. Neuroscience 19: 9281-9288

Ono H, Mishima A, Ono S, Fukuda H, Vasko MR (1991) Inhibitory effects of clonidine and tizanidine on release of substance P from slices of rat spinal cord and antagonism by α-adrenergic receptor antagonists. Neuropharmacology 30, 585-589

Ossipov MH, Suarez LJ, Spaulding TC (1989) Antinociceptive interactions between α_2-adrenergic and opiate agonists at the spinal level in rodents. Anesth. Analg. 68: 194-200

Packer M (1992) The neurohumoral hypothesis: a theory to explain the mechanism of disease progression in heart failure. J Am Coll Cardiol 20: 248-254

Pepperl DJ, Regan JW (1993) Selective coupling of α_2-adrenergic receptor subtypes to cAMP-dependent reporter gene expression in transiently transfected JEG-3 cells. Mol Pharmacol 44: 802-809

Pierce KL, Maudsley S, Daaka Y, Luttrell LM, Lefkowitz RJ (2000) Role of endocytosis in the activation of the extracellular signal-regulated kinase cascade by sequestering and nonsequestering G protein-coupled receptors. Proc Natl Acad Sci 97: 1489-1494

Reuter H (1996) Diversity and function of presynaptic calcium channels in the brain. Curr Op Neurobiol 6: 331-337

Riekkinen PJ, Sirviö J, Riekkinen M, Lammintausta R, Riekkinen P (1992) Atipamezole, an α_2 antagonist, stabilizes age-related high-voltage spindle and passive avoidance defects. Pharmacol Biochem Behav 41: 611-614

Rohrer DK (1998) Physiological consequences of β-adrenergic receptor disruption. J Mol Med 76: 764-772

Rohrer DK (2000) Targeted disruption of adrenergic receptor genes. Methods Mol Biol 126: 259-277

Rohrer DK, Kobilka BK (1998) G protein-coupled receptors: functional and mechanistic insights through altered gene expression. Physiol Rev 78: 35-52

Rosin DL, Talley EM, Lee A, Stornetta RL, Gaylinn BD, Guyenet PG, Lynch KR (1996) Distribution of α_{2C}-adrenergic receptor like immunoreactivity in the rat central nervous system. J Comp Neurol 372: 135-165

Ruffolo RR, Nichols AJ, Stadel JM, Hieble JP (1993) Pharmacologic and therapeutic applications of α_2-adrenoceptor subtypes. Ann Rev Pharmacol Toxicol 32: 243-279

Rump CL, Bohmann C, Schaible U, Schöllhorn J, Limberger N (1995) α_{2C}-Adrenoceptor-modulated release of noradrenaline in human right atrium. Br J Pharmacol 116: 2617-2624

Sallinen J, Haapalinna A, Viitamaa T, Kobilka BK, Scheinin M (1998) Adrenergic α_{2C}-receptors modulate the acoustic startle reflex, prepulse inhibition, and aggression in mice. J. Neuroscience 18: 3035-3042

Sallinen J, Haapalinna A, Viitamaa T, Kobilka BK, Scheinin M (1998) D-amphetamine and L-5-hydroxytryptophan-induced behaviours in mice with genetically-altered expression of the α_{2C}-adrenergic receptor subtype. Neuroscience 86: 959-965

Sallinen J, Link RE, Haapalinna A, Viitamaa T, Kulatunga M, Sjoholm B, Macdonald E, Pelto-Huikko M, Leino T, Barsh GS, Kobilka BK, Scheinin M (1997) Genetic alteration of α_{2C}-adrenoceptor expression in mice: influence on locomotor, hypothermic, and neurochemical effects of dexmedetomidine, a subtype-nonselective α_2-adrenoceptor agonist. Mol Pharmacol 51: 36-46

Saunders C, Limbird LE (1999) Localization and trafficking of α_2-adrenergic receptor subtypes in cells and tissues. Pharmacol Ther 84, 193-205

Scheinin M, Lomasney JW, Hayden-Hixson DM, Schambra UB, Caron MG, Lefkowitz RJ, Fremeau RTJ (1994) Distribution of α_2-adrenergic receptor subtype gene expression in rat brain. Brain Res Mol Brain Res 21: 133-149

Schramm NL, Limbird LE (1999) Stimulation of mitogen-activated protein kinase by G protein-coupled α_2-adrenergic receptors does not require agonist-elicited endocytosis. J Biol Chem 274: 24935-24940

Starke K, Endo T, Taube HD (1975) Pre- and postsynaptic components in effect of drugs with α adrenoceptor affinity. Nature 254, 440-441

Starke K, Göthert M, Kilbinger H (1989) Modulation of neurotransmitter release by presynaptic autoreceptors. Physiol Rev 69: 864-989

Stone LS, Broberger C, Vulchanova L, Wilcox GL, Hökfelt T, Riedl MS, Elde R (1998) Differential distribution of α_{2A} and α_{2C} adrenergic receptor immunoreactivity in the rat spinal cord. J. Neuroscience 18: 5928-5937

Stone LS, MacMillan LB, Kitto KF, Limbird LE, Wilcox GL (1997) The α_{2A} adrenergic receptor subtype mediates spinal analgesia evoked by α_2 agonists and is necessary for spinal adrenergic-opioid synergy. J Neurosci 17: 7157-7165

Stader CD, Sigal IS, Register RB, Candelore MR, Rands E, Dixon RA (1987) Identification of residues required for ligand binding to the β-adrenergic receptor. Proc Natl Acad Sci U S A 84: 4384-4388

Sullivan AF, Dashwood MR, Dickenson AH (1987) α_2 adrenoceptor modulation of nociception in rat spinal cord: location, effects and interaction with morphine. Eur J Pharmacol 138: 169-177

Surprenant A, Horstman DA, Akbarali H, Limbird LE (1992) A point mutation of the α_2-adrenoceptor that blocks coupling to potassium but not to calcium currents. Science 257: 977-980

Takano Y, Yaksh TL (1992) Characterization of the pharmacology of intrathecally administered α_2 agonists and antagonists in rats. J Pharmacol Exp Ther 261: 764-772

Talley EM, Rosin DL, Lee A, Guyenet PG, Lynch KR (1996) Distribution of α_{2A}-adrenergic receptor-like immunoreactivity in the rat central nervous system. J Comp Neurol 372: 111-134

Trendelenburg AU, Hein L, Gaiser EG, Starke K (1999) Occurrence, pharmacology and function of presynaptic α_2-autoreceptors in $\alpha_{2A/D}$-adrenoceptor-deficient mice. Naunyn-Schmiedeberg's Arch Pharmacol 360: 540-551

Trendelenburg AU, Limberger N, Starke K (1993) Presynaptic α_2-autoreceptors in brain cortex: α_{2D} in the rat and α_{2A} in the rabbit. Naunyn-Schmiedeberg's Arch Pharmacol 348: 35-45

Uhlen S, Lindblom J, Johnson A, Wikberg JE (1997) Autoradiographic studies of central α_{2A}- and α_{2C}-adrenoceptors in the rat using [^{3}H]MK912 and subtype-selective drugs. Brain Res 770: 261-266

Wang CD, Buck MA, Fraser CM (1991) Site-directed mutagenesis of α_{2A}-adrenergic receptors: identification of amino acids involved in ligand binding and receptor activation by agonists. Mol Pharmacol 40, 168-179

Wang R, MacMillan LB, Fremeau RT Jr, Magnuson MA, Lindner J, Limbird LE (1996) Expression of α_2-adrenergic receptor subtypes in the mouse brain: evaluation of spatial and temporal information imparted by 3 kb of 5' regulatory sequence for the α_{2A} AR-receptor gene in transgenic animals. Neuroscience 74: 199-218

Wang RX, Limbird LE (1997) Distribution of mRNA encoding three α_2-adrenergic receptor subtypes in the developing mouse embryo suggest a role for the α_{2A} subtype in apoptosis. Mol Pharmacol 52: 1071-1080

Ward RJ, Milligan G (1999) An Asp79Asn mutation of the α_{2A}-adrenoceptor interferes equally with agonist activation of individual $G_i\alpha$-family G protein subtypes. FEBS Lett 462: 459-463

Waterman SA (1997) Role of N-, P- and Q-type voltage-gated calcium channels in transmitter release from sympathetic neurones in the mouse isolated vas deferens. Br J Pharmacol 120: 393-398

Westfall TC (1977) Local regulation of adrenergic neurotransmission. Physiol Rev 57, 659-728

Wilcox GL, Carlsson KH, Jochim A, Jurna I (1987) Mutual potentiation of antinociceptive effects of morphine and clonidine in rat spinal cord. Brain Res 405: 84-93

Wilson MH, Limbird LE (2000) Mechanisms regulating the cell surface residence time of the α_{2A}-adrenergic receptor. Biochem 39: 693-700

Wise A, Watson-Koken MA, Rees S, Lee M, Milligan G (1997) Interactions of the α_{2A}-adrenoceptor with multiple G_i-family proteins: studies with pertussis toxin-resistant G-protein mutants. Biochem J 321: 721-728

Wu G, Krupnick JG, Benovic JL, Lanier SM (1997) Interaction of arrestins with intracellular domains of muscarinic and α_2-adrenergic receptors. J Biol Chem 272: 17836-17842

Yaksh TL (1985) Pharmacology of spinal adrenergic systems which modulate spinal nociceptive processing. Pharmacol Biochem Behav 22: 845-858

Zhang C, Davies MF, Guo TZ, Maze M (1999) The analgesic action of nitrous oxide is dependent on the release of norepinephrine in the dorsal horn of the spinal cord. Anesthesiology 91, 1401-1407

Zhu QM, Lesnick JD, Jasper JR, MacLennan SJ, Dillon MP, Eglen RM, Blue DRJ (1999) Cardiovascular effects of rilmenidine, moxonidine and clonidine in conscious wild-type and D79N α_{2A}-adrenoceptor transgenic mice. Br J Pharmacol 126, 1522-1530 .